CLASSICS OF BIOLOGY

CLASSICS OF BIOLOGY

BY
AUGUST PI SUÑER

AUTHORIZED ENGLISH TRANSLATION BY
CHARLES M. STERN

PHILOSOPHICAL LIBRARY
NEW YORK

Published 1955, by Philosophical Library, Inc.
15 East 40th Street, New York 16, N.Y.

TRANSLATOR'S PREFACE

THE writer of the present work, Dr. August Pi Suñer, is the sixth of a line of world-famed Catalonian biological, medical and research scientists. At the age of 24 he was successful in competition for the Chair of Physiology at Seville University, later passing to the same Chair at Barcelona University, whilst his long career since that time in the service of science covers more than half a century of great distinction and international honour. He is a Knight of the French Legion of Honour, member of the Emperor Leopold Academy at Halle an der Saale, founded by Goethe, and of the Faculties of Medicine or Science in the Universities of Catalonia, Paris, Coimbra, Genoa, Buenos Aires, and of several Spanish Universities. He was for a considerable time President of the Academy of Medicine in Barcelona, Honorary or Active Member of the Biological Societies in Belgium, Spain, Paris, Barcelona, Buenos Aires and New York, Doctor *honoris causâ* of the Universities of Toulouse and Caracas, of the Medical School of Montevideo, and holder of many other distinctions. He has taken part in numerous important international commissions and scientific congresses in all parts of the world, including Great Britain and the United States. His tall, well-built, finely preserved figure and impressive voice have indeed been a constant feature at all International Physiological Congresses since the Heidelberg meeting in 1907 down to the latest at Montreal in 1953.

Dr. Pi Suñer went to Venezuela in 1939 and was immediately asked to establish a course of Physiopathology in the Central University of Caracas, thus presenting to professional men and students in that part of South America their first opportunity to attend on their home ground a brilliant series of lectures by an unrivalled authority of universal fame in this branch of knowledge. Shortly afterwards, he was entrusted with the task of creating the Institute of Experimental Medicine of Caracas University, and of putting it on a sound scientific basis to assure continued progress. The success of his undertaking is witnessed by the fact that, though small in its beginnings, this Institute now enjoys throughout South America a fame and reputation as international as that of its founder.

Among the multitudinous writings of this distinguished physician

v

and biologist, we may mention a host of papers on his researches, for
instance—

> *The Origin of the Bile Pigments* (1903)
>
> *Renal Functions and Urinary Secretion* (1905–47)
>
> *Respiratory Chemoreceptors* (1916–47)
>
> *Trophic and Visceral Reflexes* (1917–41)
>
> *The Physiological Mechanism of Natural Immunity* (1904–5), in
> collaboration with R. Turró

and of his lengthier books—

> *Anaerobic Life* (1900)
>
> *The Functional Unit* (1919)
>
> *The Mechanisms of Physiological Correlation* (1920)
>
> *Trophic Sensitivity* (1941)
>
> *Dispersed and Connected* (1945)
>
> *The Vegetative Nervous System* (1947), Prix Pourat, Paris, 1948
>
> *The Bridge of Life* (1951)
>
> *Classics of Biology* (1954)

All these works first appeared in Spanish and many have since been
translated into French and other languages. The present book is the
first translation into English of the last title on the foregoing list, and
every effort has been made to give a faithful reflection of Dr. Pi Suñer's
clear and forceful style.

The translator has enthusiastically taken the opportunity of
returning to the originals of the classics selected by Dr. Pi Suñer
to illustrate each subject and of rendering them afresh into English.
Where the classic was originally in English, of course, the original text
itself is quoted and thanks are here expressed for the ever courteous
help given by the staff of the Printed Books Department of the British
Museum in tracing basic sources and for the facilities afforded by the
National Central Library Service.

C. M. S.

CONTENTS

vii

Contents

x

Contents

CHAPTER I

MATTER AND ENERGY IN LIFE

AGENT AND ACTION

PRIMITIVE man dissociates his idea of motive agent or impelling instrument from that of the body set in motion. Direct observation in fact teaches us that for a body to start moving some force must be applied to it. This idea was first scientifically stated by Aristotle and upon it was founded the rational mechanics of Galileo and Newton. Man does not intuitively perceive that movement may be but one of many properties of matter, inherent in the manner in which such matter exists.

Animate beings are indeed distinguished through their very capacity for motion. An animal moves and gives off heat or even light in some form; every living body is capable of growth. Not so very long ago, in 1907, Strecker in his *Der Kausalitätsprinzip in der Biologie* (Cause and Effect in Biology) was still able to say: "Every organic body is a form of activity, whilst a purely chemical or physical object can do nothing more than be set in motion." Since it is impossible for a thing to become animated by itself alone, some motive principle special to living matter must be in existence.

Throughout the ages Biology has been vitalistic. Looking back through the history of physiological knowledge we see the continuous passing-on of the idea of a breath, spirit, ether or animating principle acting upon pure matter, that is, upon a passive immovable substance possessed of weight but incapable of action by itself alone. In the course of time these ideas have taken many different forms, though all are essentially similar.

Whence springs the activity present in live organisms? One of the most important steps in the progress of Biology was signalized by proof that the physical laws governing living matter are the same as those applying to the inorganic world.

COMBUSTION

It is essential to take an objective view of matter and of the Universe in general. Certain bodies burn up. The Greek philosophers, Aristotle for one, considered fire to be one element out of the four—air, earth, fire and water—which, among them, covered the whole field of existence as then known. Combustion was held to be the liberation of the element *fire* from ignited bodies. According to the alchemists this element contained sulphur, or as they called it, brimstone. Becher in 1699 and Stahl in 1702 suggested a similar explanation for combustion, namely, that a hypothetical substance, which they called *phlogiston*, escaped from burning bodies and diffused into the air wherein it disappeared completely.

Subsequent to Scheele (1771), Priestley (1774) recognized the fact that air did contain oxygen (previously isolated by Borch in 1678 and Hales in 1729 though it had not hitherto been given a name). Priestley showed that oxygen aids combustion and interpreted the phenomenon on the assumption that such oxygen was *dephlogisticated* air which favoured the absorption of phlogiston.

Had this theory been correct, then the products of combustion should have weighed less than at the start of the process, since phlogiston is supposed to be liberated. Actually the opposite takes place, as in all cases of oxidation. Lomonossov in 1756 put a quantity of tin in a hermetically sealed flask and heated it up until the tin turned to a white powder. On opening the flask, air rushed in, thus proving that a part of what had originally been in the flask had disappeared. The weight of the white powder was greater than that of the original tin, so something must have combined with the metal—something which had previously been present in the air. Lavoisier in 1775 confirmed these facts and gave a definite explanation thereof. He heated quicksilver for four days inside a closed retort, at a temperature slightly below boiling point, and a red crust then formed on the surface of the metal. This product on being heated still more intensely separated into quicksilver, which could be recovered, and a gas. The volume of such gas equalled that lost by the air in the first part of the process and its properties were identical with those of the gas Priestley obtained when using mercuric oxide as his starting point. The weight of the product obtained on heating mercury was greater than that of the original quantity—exactly as Lomonossov had observed—and the increase tallied with the weight of gas disappearing from the air enclosed in the retort and which was given off when the oxide had disintegrated. The oxidation process must consist therefore in fixing one of the components of air on to a definite quantity of a substance, and Lavoisier decided to call such component *oxygen*, that is, *acid producer*.

Lavoisier founded modern chemistry by the use of the balance, following up the conversion of substances by studying their weights. He established the Law of the Conservation of Matter, which states that the properties of substances vary as they react upon one another but that the total quantity of matter remains unaltered. Nothing is lost and nothing gained in a reaction or in a series of reactions. Combustion is therefore, in general, a form of oxidation. The burning of coal in a grate is the same as oxidizing mercury, for instance. The laws governing each process are the same. Carbon dioxide, the product of the combustion of coal, weighs more than the coal itself, and the increase in weight is exactly equal to that of the oxygen consumed. Ideas on combustion at this stage have stridden a long way ahead of the old phlogistic theory.

RESPIRATION

Let us look back once more. In 1667, Robert Hooke demonstrated to the Royal Society the need for "fresh air"—these are his actual words—"for maintaining life." Mayow in 1674 proved that the component of air necessary for breathing purposes was the *nitro-aerial spirit*, oxygen, which had already been found to be the substance essential for promoting combustion. He assumed, furthermore, that such nitro-aerial spirit must penetrate into the blood where it makes itself indispensable to the vital activities, by combining with the sulphurous particles in the blood in the same way as would happen in burning the substance.

In 1774, Priestley noticed that if mice were left for some time in a closed chamber, continuance of life was impossible because the air became impure. Living plants introduced into the chamber made the air breathable once more. Soon afterwards, Jan Ingenhousz in 1780 discovered that this occurred only if the plants were exposed to light: "The sun's light induces in the leaves that activity which generates dephlogisticated air." Black in 1755 had proved that animals produced, when breathing, a gas different from air, which he called *fixed air*. Lavoisier soon showed that this gas is the same as that obtained when coal is burnt, namely carbon dioxide. Jean Senebier in 1780 gave the explanation of the occurrences noted by Priestley and Ingenhousz, thus: "The leaves of plants convert fixed air into dephlogisticated air," that is, they turn carbonic gas into oxygen. A long time was still to elapse before the isolation and discovery of the properties of chlorophyll by Pelletier and Caventon in 1818!

Lavoisier arrived at systematized and precise concepts regarding respiration. Beings living in air consume oxygen and exhale carbon dioxide in practically equal amounts. Combustion takes place within the organism and, as in the physical world, heat is generated. Lavoisier resorted to Laplace to get some idea regarding the means necessary to enable such produced warmth to be measured. He made the first calorimeter for biological use and completed the first calorimetric determinations ever carried out on living matter.

CONVERSION OF ENERGY

Several years of political convulsion were to pass in Europe, the Napoleonic Wars with all their various consequences, and thus, nothing of note came to light in our field until 1824, when Sadi Carnot published his celebrated work *Réflexions sur la Puissance Motrice du Feu* (Reflections on the Motive Power of Fire). In due course there had arisen the peculiar paradox that though the steam engine was becoming more important because of its enormous value in industry, its invention had come about from purely empirical observation. It was of supreme interest to know the principles in virtue whereof such extraordinarily potent results could be obtained. Carnot studied the physical conditions obtaining in steam engines at work and discovered the so-called Second Law of Thermodynamics, namely that a difference of temperature occurs in various parts of an engine and that heat must pass from a warmer to a cooler portion, like water falling to a lower level and turning a millwheel. Heat in this case thus behaves like a material element.

In spite of the importance of Carnot's law, light did not yet break openly on the subject. Some twenty years later, an obscure German doctor, Julius Robert Mayer, wrote *Die Mechanik der Wärme* (Mechanics of Heat) in 1841, but his thesis did not achieve publication as it conflicted with the ideas then reigning in the world of physics. Among his points was the following statement: "We must bring ourselves to think of two such dissimilar items as mechanical work and heat as two forms of the same thing." Heat can be converted into work and vice versa, such conversions being just as quantitative as transformations of matter itself. In changes of energy, in physics, as in chemical reactions, there is neither gain nor loss of energy throughout the whole series of possible conversions. There is a law of the conservation of energy just as valid as that of the conservation of matter.

Very soon, Mayer's ideas were confirmed by the work of James

Prescott Joule, an amateur—like so many English scientists at the start—who in 1843 published his first thesis on heat and work.

A great scientist, the physiologist Helmholtz, in 1847 proved conclusively and with a sure hand the arguments above mentioned. He was at that time but twenty-six years old and for the moment did not possess sufficient reputation as a scientist. Like Mayer, he met with difficulty in getting his work published and his ideas accepted.

METABOLISM AND ENERGY

Mayer had already asserted that the Law of the Conservation and Conversion of Energy is applicable to the activity of live bodies. In 1845 he published his treatise *Ueber die organische Bewegung in ihrem Zusammenhang mit dem Stoffwechsel* (On Organic Motion in its relationship to Metabolism), in which he showed that from the physical point of view, vital phenomena are indistinguishable from what occurs in the inorganic world. Helmholtz definitely set forth his views in his much-discussed work *Ueber die Erhaltung der Kraft* (On the Conservation of Energy), published in 1847. A group of young physicists, mathematicians and physiologists had founded The Berlin Physical Faculty, and whilst Helmholtz's reception on the part of the older diehards of the Academy of Sciences had been extremely cold, he aroused in the younger group such an enthusiasm as has not been quenched even to this day.

Complementary to these discoveries was the codification of chemical and thermochemical energy. Marcellin Berthelot in 1875 wrote his celebrated treatise demonstrating that mechanical energy is reproduced in chemical processes. In any chemical reaction conversions of matter occur simultaneously with displacements of energy. In this way exothermic or endothermic reactions are developed.

The calorimetric measurements taken in live bodies by Berthelot himself, Pettenkoffer, Atwater, Benedict, Du Bois, and many other physiologists, make it quite plain that the laws of thermochemistry govern nutritional processes in animals and plants just as they do in inorganic bodies, and that assimilation is a combination of synthetic reactions whose energy balance is endothermic, as for instance in the great assimilative mechanism represented by the action of chlorophyll, where energy supplied by the light of the sun is effectively utilized.

On the other hand, disassimilation in its more distinct forms is an exothermic breaking-down or release of heat and this energy appears in

life under many different guises, such as motion, warmth, electric
power, light, syntheses coupled with disassimilative reactions, and the
rest. In the living world the laws of the conservation of matter and of
energy recur. Energy in living beings is a special manifestation of the
universal energy of the Cosmos.

Let us return to Lavoisier who was the first to express the whole thing
with clarity. When coal burns in a stove, fixing oxygen and liberating
carbon dioxide, such chemical change gives out heat, namely the heat
of the fire. Animals and plants breathe, consuming oxygen and
exhaling carbon dioxide gas. Combustion of carbon takes place
within their organism, as well as many other biochemical processes and
exothermic disassimilative reactions. Here also the combustion of
carbon, just the same as in the stove, sets a certain amount of energy at
liberty. The same thing occurs in the other reactions taking place
at the same time in the living creature, and such energy appears in the
form of heat and in other shapes such as activity in the functions, or
other living processes. Looked at from the energy side of things, every
living organism is a thermochemical machine.

*   *   *   *   *

ANTOINE-LAURENT LAVOISIER 1743–94

*Mémoire sur la nature du Principe qui se combine avec les Métaux pendant
leur calcination et qui en augmente le poids* (On the Nature of the
Principle which combines with Metals in the course of Burning and
increases their Weight), Paris: *Mémoires de l'Académie Royale des
Sciences,* 1775.

The substance which combines with metals subjected to heat and
increases the weight thereof is none other than the purest portion of
the air surrounding us and breathed in by us, and which by such
process passes from a state of flexibility to that of fixation.

When charcoal is burnt, it disappears in gaseous form, but the
resultant gas is greater in weight than the original charcoal consumed,
because this latter has combined with the pure portion of the air.
This is identical with what happens when quicksilver is burnt. Conversely, a strong heating-up of the burnt product allows quicksilver to
precipitate and air is then given forth which can be breathed freely.

When charcoal is used for reducing metals such as quicksilver and the rest, it completely disappears whilst the metal returns to its original state, fixed air (carbon dioxide) being generated at the same time. From this we may deduce that the substance which has been called *fixed air* is a combination of the pure substance, breathable oxygen from the air, with charcoal. I propose to analyse this conclusion in detail in the following series of notes.

EXPERIMENTS ON THE BREATHING OF ANIMALS

Among all the phenomena of animal economy none is more striking nor more worthy of attention on the part of philosophers and physiologists, than those occurring in the act of breathing. We know very little in regard to this function; merely that it is necessary for the preservation of life and cannot therefore be suspended for any length of time without exposing living creatures to the risk of instant death.

The experiments carried out by certain philosophers, especially Hales and Cigna, have thrown a certain light in this connection. Doctor Priestley in an important work has recently widened the range of our knowledge. He has tried to show by means of certain extremely ingenious, novel, and delicate experiments, that animal breathing has the property of phlogisticating the air in a way similar to that seen in the combustion of certain substances and in other analogous chemical processes. The air then becomes unfit for breathing purposes and this occurs when it is overloaded and eventually saturated with phlogiston.

However plausible the theory of this celebrated philosopher may appear to be at first sight, and however well conducted may have been the large number of experiments upon which their author relies, I must confess that I find some of the phenomena described to be so mutually contradictory as to give rise to certain doubts on my part. Consequently, I have myself experimented by different methods and the results of my researches have led me to conclusions distinctly at variance with his.

Air which has been used for the combustion of metals contains nothing more than the noxious residue from atmospheric air. The part thereof formerly fit for breathing becomes combined with quicksilver during combustion. The air breathed in, once separated from the fixed air, is exactly the same as that entering into combustion. In fact, by combining with the breathable residue, somewhere about a quarter of its volume of dephlogisticated air (oxygen) extracted from the mercury consumed, I was able to restore the air to its original state perfectly suitable for respiration, combustion, etc. The same occurs if air vitiated by combustion of mercury be used. The addition of a proper amount of the pure principle obtained by long and intense heating of the said calcinated mercury, restores to the air its qualities of breathability and its capacity for combustion.

From these experiments it is evident that in order to restore air which has been rendered noxious by breathing, we must—

Firstly, get rid of the fixed air contained therein, using quicklime or caustic alkali for such purpose, and

Secondly, add an amount of pure breathable or dephlogisticated air equal to what has been lost. Respiration gives the converse effect.

In fact, in my paper read at the Public Session last Easter, I pointed out that dephlogisticated air can be converted into fixed air by burning powdered charcoal in it, and that there are various other means of obtaining a similar result. It is possible that respiration possesses the same property, namely that dephlogisticated air taken in by the lungs may be returned in the form of fixed air. Pure air may combine with the blood and that might be the reason for the red colour distinguishing the latter.

We may take it for granted—

1. Breathing changes only the pure or dephlogisticated air contained in the atmosphere. The residue or noxious portion enters the lungs and is expelled in the same state without change or modification.

2. The burning of metals in a given quantity of atmospheric air takes place proportionately to the dephlogisticated air contained therein, and combining with the metal.

3. Similarly, an animal confined in a given quantity of air which is not renewed, will die after having consumed the breathable portion of such air, and converting it thus into fixed air which is useless for the purpose of breathing.

4. The noxious air, not fit for breathing, remaining from the calcination of metals is identical to what remains from respiration, once the fixed air has been separated.

The noxious air in one case or the other can be restored to the atmospheric state if a quantity of dephlogisticated air equal to what disappeared in the burning or breathing process be incorporated therewith.

MARCELLIN PIERRE-EUGÈNE BERTHELOT 1827–1907

La Synthèse Chimique (Chemical Synthesis), Paris: Gauthier-Villars, 1875.

Initial attempts to analyse organic substances led to results so strange that the problem itself, approached with certain tenacity from the seventeenth century to the beginning of the eighteenth, seemed to many at that time to be impossible of solution.

Later researches showed, however, that the component parts of animals and plants, like the products deriving therefrom, are the only substances capable of giving oils and products of combustion on distillation. This feature distinguishes them from mineral compounds which never give rise to such products of combustion. Thus, once for all, the distinction between mineral chemistry and organic chemistry was established.

Furthermore the way was thereby paved for the discovery of the constituent elements of animal and plant matter. It had been observed that all plants are formed from the same elements, since they all give similar products on decomposition, namely water, oils, carbonaceous earths and the rest. The same applies with even more force to animals, as animal elements are up to a point identical with those found in plants. Indeed, the decomposition of animal matter gives substances identical with those obtained from the decomposition or distillation of plants, with the addition of considerable proportions of volatile alkalis, all of which seems to indicate some factor characteristic to animal chemistry. The presence of such volatile alkali is precisely the thing which from the chemist's viewpoint distinguishes animal matter.

When later the nature of water, of oils produced by heat, and of the volatile alkali became known and it was demonstrated that all these bodies were combinations of single elements, it was possible to deduce that carbon, hydrogen, oxygen and nitrogen are the four basic elements of living bodies.

Here we have an all-important result which suffices to show that these analytical studies, for all their inevitable imperfections, were of real value. Thanks to work on the chemistry of the air, interpretation of these facts suddenly acquired greater clarity.

The discovery of the gases, the theory of combustion, the proof of the constancy in weight of bodies in chemical reactions, and of the stability of single bodies or elements throughout the whole range of conversion—in short, the sum total of knowledge definitely obtained by Lavoisier and his contemporaries—placed the science of chemistry on the solid foundation on which it still stands today. They gave a positive analytical basis to organic chemistry, as had previously been the case with mineral chemistry.

It was now possible to show that plants are formed mainly from carbon, hydrogen and oxygen and that animals contain these same elements with the additional feature of considerable amounts of nitrogen. Such simplicity in the chemical composition of live beings is so much the more interesting since the same elements are also met with in inorganic mineral compounds.

Towards the middle of the eighteenth century, ideas regarding the composition of organic substances became somewhat more precise. Taking into account the differences observed between the properties of products from analysis and those of the animal or plant substances analysed, chemists began to concern themselves with the question of isolating in their natural state not only the elements but also the "immediate principles" which made up the substance of living bodies.

A fair number of these immediate principles were isolated, recognized and identified in the same century. Certain acids, such as tartaric, oxalic, malic, citric, lactic, uric, formic, gallic, benzoic and the rest, and other substances like urea, sugar of milk, cane sugar, cholesterine or wax from the biliary calculi, not possessing the property of acidity, were labelled. The foundation of organic chemistry, as M. Chevreul says, is the exact definition of the species of immediate principles in animals and plants. Every research into living phenomena requires an exact knowledge of the immediate principles which make up the tissues and liquids within the organism, where the phenomena in question occur.

Reproduced by permission of the publishers

HERMANN LUDWIG FERDINAND VON HELMHOLTZ
1821–94

Ueber die Erhaltung der Kraft (The Conservation of Energy), Berlin: Reimer, 1847.

Since the present discussion had from its nature to be addressed essentially to physicists, I have chosen to present its principles purely in the form of a physical hypothesis independently of any philosophical argument, developing their consequences and comparing these with the empirical laws governing natural phenomena in the various branches of physics. The derivation of the principles established can be set about from two points of view, namely, the postulate that it is impossible to acquire unlimited mechanical power through the action of any one combination of natural bodies on another, or the assumption that all activities in nature may be attributed to forces of attraction and repulsion, whose intensity depends solely on the distance between their points of application. That both these theories are identical is demonstrated right at the very start of this discussion. However, they still have an essential significance for the final and real aim of the physical sciences in general, which I shall endeavour to set forth in this separate introduction.

The problem of science is to discover once for all the laws whereby individual events in nature may be traced back to general laws and be in turn determined from these latter. Such laws, for instance, as those of the refraction or reflection of light, Boyle's and Gay-Lussac's laws for the volume of gases, are obviously nothing but general notions, by means of which all phenomena appertaining thereto are gathered together. The search for these is the business of the practical side of our sciences. Theoretical science, on the other hand, seeks to discover the unknown origins of the processes by considering their visible effects, seeking to comprehend them by application of the principle of cause and effect. We are led to and entitled to such practice through the axiom that every change in nature must have its own adequate cause. The proximate causes, which we attribute to natural phenomena, can well be invariable or variable; in the latter case, the same principle compels us to seek in their turn other antecedent causes for such modification, and so on, until we eventually arrive at ultimate causes operating in accordance with an immutable law which consequently at all times produces the same effect under the same external conditions.

The ultimate aim of pure science is therefore to seek out the final unalterable causes of the processes of nature. If indeed all processes are to be traced back to such causes, if therefore nature is bound to be entirely comprehensible, or if there be modifications in it which elude the law of compulsory causality, and belong therefore to the sphere of free chance, this is not the place to discuss such possibility; in any event, it is clear that science, whose mission it is to understand nature, must proceed on the assumption of its being intelligible and according to such intelligibility make inquiry and inference until becoming compelled, perhaps by irrefutable facts, to acknowledge its own limitations.

Science considers the affairs of the universe according to two abstract principles, in the one case from their mere existence apart from their effects on each other or on our organs of perception, designating them in such case as *matter*. The existence of matter in the abstract is to us therefore something passive and inactive; we discern in matter its distribution in space and its mass, which are laid down as being constant in perpetuity. We should not ascribe qualitative differences to matter as such, for if we speak of heterogeneous substances, we then fix their dissimilarities only all the more in the dissimilarity of their effects, that is in the forces they exert. Matter in itself cannot, for that reason, admit the slightest change other than a spatial one which is *movement*. Objects in nature are not, however, inactive, for we generally arrive at our knowledge of them only through the activities impinging from them on to our perceptive organs, whilst from such effects we infer an effective cause.

If we want, therefore, to apply the notion of matter to reality, we may do so only by adding thereto still a second abstract idea, which in the first instance we were seeking to exclude: to wit, the ability to exert effects, which is the same thing as ascribing forces to such effects. It is clear that the two ideas of matter and force applied in nature may never be separated. Pure matter would be of little moment for the rest of nature, since it could never establish a modification therein or in our organs of perception; a pure force would be something which ought to exist and yet on the other hand have no actual existence, for "matter" is the name we give to what is existent. It is equally erroneous to seek to define matter as something concrete, and force as a mere idea not corresponding to anything actual; both are, on the contrary, abstractions, each formed in exactly the same way from reality; we can in fact only perceive matter from the very action of its forces, and never in the abstract.

We have seen above that natural phenomena should be taken back to immutable ultimate causes; this requirement thus turns out in such form that in the temporal sense invariant forces have to be found as ultimate causes. Substances with invariant forces (indestructible qualities) have been defined by science as (chemical) elements. If however we consider the Universe to be composed of elements possessing invariant qualities, then the only changes still possible in such a system must be in spatial position, which is to say movement, and the external conditions by which the effect of forces becomes modified can themselves only be spatial in their turn, so that forces are only forces of motion or impetus, dependent for their action only upon spatial relationships.

Or, to put it more clearly, natural manifestations have to be *reduced* to movements of matter whose *fixed* powers of motion depend solely on *spatial circumstances*.

Motion is a modification of relative positions in space. Spatial relationships are possible only in comparison with *bounded quantities* of space and not as against *undifferentiable* empty nothingness. Motion can therefore occur only in practice as a change in relative position of at least two material bodies in relation to each other; kinetic energy as its cause *will therefore also only be inferable* from a comparison of at least two bodies as against each other and is thus to be defined as the effort of two masses to change their position in relation to each other. The force, however, which two integral masses exert on each other must be *resolved* into the individual forces of each of their material particles, i.e. of their molecules. Such points of space filled with matter have, however, no spatial connection with each other apart from the distance between them, since the direction of their line of union can be defined only from a comparison with at least two other points. Any impulse, which they exert on one another, can therefore give rise only to a change in the distance between them and is therefore a force of attraction or of repulsion. This also *validly* follows directly from theory. The forces which two masses exert on each other must necessarily be determined by their size and direction immediately the position of the masses is completely defined. However, only one single direction is completely determinable through two points, *to wit*, that of the *line joining their centres of gravity;* consequently the forces they exert on each other must lie along this line and their intensity can depend only upon the distance between them.

Thus, the ultimate task of the physical sciences is to *reduce* natural phenomena to invariant forces of attraction and repulsion whose intensity is dependent on their distances apart.

The resolution of this problem is at the same time the necessary condition for a complete understanding of Nature. Pure mechanics has not hitherto *admitted* this limitation for the concept of kinetic energy, *firstly* because it was not itself clear as to the source of its principles, and secondly because it is *important* to be able to calculate the *results* of complex forces in those cases where their *resolution* into *simple* ones has not yet been attained. Moreover, a considerable portion of the general principles of motion of compound systems is demonstrable only in the case when these act on each other through fixed forces of attraction or repulsion; that is, the law of virtual velocities, the *conservation* of the *plane of the principal axis of rotation* and of the moment of rotation of unconstrained systems, the principle of the conservation of energy. Only the first and the last of these principles are in the main applicable under terrestrial conditions, since the others relate only to completely free systems; the first again, as we shall show, is a special case of the latter, which therefore appears to be the most general and most important consequence of the *deductions drawn*.

Pure physics, therefore, if it is not to stick half way along the road to *intelligibility* must arrange its *views*, on the nature of the simple forces and their consequences, in harmony with the challenge laid down. Its task will only be accomplished when the reduction of phenomena into simple forces has finally been consummated and when it can at the same time be shown that this is the only possible deduction permissible under the circumstances. This then might prove to be the essential concept of the constitution of nature, to which the quality of objective truth would also thus be ascribable.

WILHELM OSTWALD 1853–1932

Die Energie (Energy), Leipzig: Barth, 1908.

All living organisms build up their energetic system in the first place upon the basis of chemical energy. The reason for this is that chemical energy is the most concentrated and most conservable of all the forms of energy available. We have already observed that the stock of free energy, to which all living creatures owe their existence, is being constantly renewed and delivered to us from the sun. The sun, however, only shines half the day on the average and stays beneath the horizon for the other half. For any organism, therefore, to enjoy a *permanent* supply of such energy, there must be some kind of arrangement whereby it may not only absorb a proper amount for its immediate needs but garner in also a sufficient stock to take care of its existence for the other half of the day. For, if it so happened that the constant flow of energy through the unchanging organism, which is the very essence of its life, were interrupted for any appreciable period, then the machine would as a rule be absolutely unable to set itself going again, even though free energy were once more to become available. As a matter of fact it becomes increasingly evident in the preponderating majority of organic beings that even a very transitory stoppage may be sufficient to bring things to a permanent standstill and thus cause death.

Every provision must therefore be made to convert the daily supply of energy into some suitable permanent form for preserving the flow at least during the night, and even throughout the winter also, without the creature being reduced to extremities. Of all the forms of energy known to us, none fulfils this requirement better than chemical energy. We see this from the fact that chemical energy serves us as the stock for all other purposes of existence and industry. Our foodstuffs consist of chemical energy, and the great source of energy for all industry up to the present has been almost exclusively bituminous coal reconverted into chemical energy. It is only quite recently that we have begun to use to any great extent the mechanical energy in the form of elevated masses of water put at our disposal by the sun.

Owing to the general state of affairs on this earth, the constitution of our organic life depends even to its smallest details on the necessity of founding the energetic economy of organic beings on the chemical form of energy. Our muscles work on chemical energy, and similarly the working of our nerves, still something of a mystery to us, is closely bound up with this form of energy. In particular, however, a special property of all organic phenomena depends also in all probability on chemical relationships, and that is the phenomenon of *consciousness* or intelligence in the most general sense, as was first discovered by E. Hering.

For instance, organic beings frequently demonstrate the property of being able to repeat with much greater ease such processes as have already happened to them one or more times than to carry out something which is entirely new to them. This is by no means a general thing in nature, and is in fact so rare that in the inorganic world the contrary is more often the rule. An electric cable does not carry the current any the more efficiently just because it has already served for that purpose, any more than a kettle comes more rapidly to the boil because it may already have been heated up some thousands of times. In other words, inorganic structures do not retain the least vestige of *previous experience* or history within themselves; a repeated occurrence is for them ever just as new an experience as it was the first time. Only in exceptional and fairly complicated cases does something of a similar nature become evident, as for instance, when a machine grinds down its moving parts through wear and tear and runs thereafter more smoothly than at the beginning. But even in a case like this, we could reduce such a running-in period to zero, if we polished the working parts beforehand to maximum smoothness. It is a question, therefore, of chance rather than intrinsic similarity.

On the other hand, such behaviour is common in the organic world. Reference has previously been made to the remarkable fact that children become extremely like their parents, even when removed from direct influence of the latter for the greater part of their adolescent years. Among the possibilities as to how parental characteristics are transmitted, the most likely is the *chemical* theory, that is the hypothesis that such peculiarities are caused by corresponding peculiarities in the germ substance. For we call to mind that, in addition to spatial energy (under which classification I include the energy of motion, gravity and conversion in volume and area), chemical energy alone has proved to be a permanent and inseparable component of ponderable matter. The boundless multiplicity of animate matter can, however, be ascribed only with difficulty to the relatively invariant energy of space; therefore, of the types of energy known, only chemical energy remains as a carrier for these diverse qualities. Of course, we do not exclude the possibility that still other forms of energy, hitherto unknown, may come into the question, though up to the present no signs whatever of these have shown themselves.

Now, chemical arrangements, possessing the same *property of recollection*, can be imagined and invented, to carry out a repeated process easier than at first shot. I am quite willing to admit that the resemblances are mainly of a purely superficial nature; yet on the other hand it must be remembered that it is but little more than a decade since the field of chemical phenomena now under consideration in this connection, namely the *catalytic* processes, has been subjected to systematic treatment, and consequently only small portions of its ramifications have meantime come to light.

It is perhaps as well at this point to make some general remarks on questions of a cognate nature. The problems of *applied* science resulting when any imminent intricate natural phenomenon has to be interpreted on the basis of our knowledge of the laws of nature, are of a particularly treacherous character. Nature obviously has at her disposal the whole stock of possibilities and expedients which are in any way applicable, whereas our attempts at interpretation have to be made with the aid of such little science as we may possess for the moment and of which nothing is more certain than that it is absolutely incomplete. Let us take but one example from recent practice: up to the present not even a relatively satisfactory idea could be found to explain the generation of heat from the sun, since the *known* sources of energy all flow too meagrely to supply the enormous needs of the sun's radiation. By the discovery of radium, which represents a source of energy some million times greater in concentration than those heretofore known, the problem has now been decidedly transferred to the realm of the explicable. It has not indeed been proved that the sun's radiation actually originates in a process of the nature of spontaneous disintegration of radium, but we no longer have any need to say that no known source exists able to meet that requirement, since we do have such sources at our disposal and may even say with some degree of probability, on the basis of this experience, that further sources may exist different in quality but similar in productivity.

Living organisms by their very activity and existence thus set us thousands of unsolved problems, amongst which many are doubtless beyond solution with the means at present at the disposal of science. The resources of science, however, expand from day to day, and what today not only seems but actually is still unintelligible may on the morrow be a very easily resolved question. Thus, the information that this or that vital phenomenon is not susceptible to interpretation not only should never be regarded as an indication that it will always remain so, but should indeed be taken as a sign of the very reverse, to the effect that—whoever seeks to interpret all these kinds of phenomena solely on the basis of today's scientific knowledge has not yet properly appreciated the basic principles of the matter. The same holds, too, for those demanding something of a like nature from the champions of scientific biology. This last is said for the ears of those who allege that the phenomena of life are basically unintelligible.

CHAPTER II

CELL THEORY

EARLY OBSERVATIONS

THE microscope was invented in 1590 by a Dutch spectacle-maker, one Zacharias Jansen, a resident of Middelburg. It was some three-quarters of a century later, before Robert Hooke was to publish in 1665 his *Micrographia* embodying the results of his researches in this field. Hooke himself was originally an architect and professor of mathematics addicted to the study of optics and interested also in what was at that time the virtually insuperable problem of building a flying machine. He produced a compound microscope which, in comparison with Jansen's and the model which Leeuwenhoek was using at about the same period, was strikingly perfect.

Hooke gives a description of the structure of cork in the eighteenth section of *Micrographia*. Cork contains, says he, numerous cavities— which he denominates "pores" or "cells"—reminiscent of the structure of a honeycomb, sponge, or pumice stone.

Almost contemporaneously, Marcello Malpighi pointed out in 1675 that cell structure is observable in many plants and in certain animal organs. For all that, neither Hooke nor Malpighi arrived at the conclusion that the basic factor was actually what was contained in the cell itself and not the material by which it was surrounded.

Around that same time, in 1673, Leeuwenhoek discovered certain free or independent cells, to which however he did not attribute any specific biological significance; these were the red blood corpuscles, spermatozoa, infusoria, yeast cells and others. Leeuwenhoek, though, was but an amateur not ranking particularly high in the opinion of his contemporaries; nevertheless his discoveries were of very real importance for the advancement of science.

These various observations remained unmentioned for more than a century because of the predilection for organistic standards in vogue at that time. The organ, both in plants and in animals, was held to be the source of all functions. At the beginning of the nineteenth century, however, Xavier Bichat in 1802 saw, in a flash of inspiration, that the organs are made up of tissues and that it is these latter which determine the physiological properties of a body and which are subject to disease.

Morphology, that is the study of structure and form, had meantime borne fruit with fresh discoveries—none, however, of startling importance. Corti asserted in 1772 that the viscous liquid which Malpighi had stated as contained in cells, was in fact an ever-present constituent thereof. Somewhat later, in 1835, Dujardin confirmed the existence of a gelatinous substance inside the cells and proposed to give it the name of *sarcode*. Johannes Purkinje in 1840 and Mohl, the botanist, in 1846 brought the name *protoplasm* into favour. In 1781 Fontana was able to give a description of a nucleus and nucleolus in cells and, in 1833, Robert Brown confirmed the existence of a small differentiated "nucleus" in plant cells and suggested that its function was of considerable importance.

THE CELL BECOMES A BIOLOGICAL UNIT

In 1826 Turpin wrote as follows: "The cell is the fundamental element in the structure of living bodies, forming both animals and plants through juxtaposition." However, the cellular theory did not take proper shape until the famous meeting between Matthias Jacob Schleiden and Theodor Schwann. The former was a lawyer who gave up his profession in order to devote himself to the study of plant structure and physiology. The latter was an anatomist specializing in zoological studies. In October, 1838, there took place a historic meeting in the course of which the two sages exchanged ideas with one another. Afterwards, Schleiden visited Schwann's laboratory and was shown certain slides of animal tissues which enabled him to verify the similarity between animal structure and plant structure; he opined that the cells are the structural units of the plant, and in the following year, 1839, Schwann asserted that the same holds also for animals. "The elemental parts of the tissues are the cells, similar in general but diverse in form and function. It can be taken for granted that the cell is the universal mainspring of development and is present in every type of organism. The essential thing in life is the formation of cells."

Later on, in 1861, Max Schultze showed that plant cells are in fact identical with those forming the lower animals. He enunciated what is now called the "Protoplasmic Theory," namely that the basic unit of organization in every living being consists of a mass of protoplasm surrounding a nucleus, and that this system of organization is common to all classes of living bodies.

CELLS IN PHYSIOLOGY AND PATHOLOGY

At the same time, the physiologists and pathologists were increasing their field of morphological knowledge. Of prime importance is Rudolph Virchow's great work *Cellular Pathology* published in 1858. "Cellular theory applied to all living bodies leads us to a cellular physiology and cellular pathology each based upon histology, that is, upon the anatomical knowledge of the elements. The seat of illness must be investigated in the cell and symptoms of disease are nothing more in the final analysis than evidence of certain reactions of the component cells of the organism in the presence of the causes of the illness." And forty years later, in 1895, he repeats: "The essence of disease is found, in my view, in some affected part of the organism, in some cell or conglomeration of cells, some tissue or some organ."

Physiologists like Johann Müller and Claude Bernard, botanists like Strasburger, recognized the truth of this theory which became universally accepted. All biologists since that time speak in terms of the cell and its components. From other quarters we see the emergence of the immortal discoveries of Louis Pasteur. Beginning with 1857 we proceed first from the discovery that the agents of fermentation are live bodies, then this idea is extended to embrace the causes of infectious diseases, which are independently living cells such as Koch was able to cultivate in 1867 outside the body itself, whilst Brücke was able definitely to formulate the general law that "every live body is a cell or a combination of cells."

Cells reproduce to form new individuals or new cells in the same individual. As Virchow wrote: "Every cell must come from a cell" (*Omnis cellula e cellula*), combating those many who since the time of Schwann held that the cell coagulated in some manner from plasmatic liquids. This is indeed how the somatic cells which form the individual exist singly or in bunches and how reproduction becomes possible by the division of specialized cells, such as the germ cells, whether sexed or not. One lives and dies by means of the cell.

Since that time cellular physiological and morphological knowledge has increased considerably. Schleicher in 1878 and Flemming in the same year were able to give a description of karyokinesis, the behaviour of nuclear elements in the mitotic division of the cell. Strasburger in 1876 notes, moreover, the centrosome as the main factor in karyokinesis.

Previously, Hertwig in 1875 had studied the fertilization process using ovule and sperm from the sea-urchin. Van Beneden in 1883 and Boveri in 1886 obtained important data whereby it is proven that fertilization is merely the combination of two cells, the male gamete with the female, which promotes the segmentation of the egg already previously observed by Rusconi in 1826 and Newport in 1853.

From all points of the compass we get fresh confirmation that the cell is the real unit of life, that living bodies may consist of one single cell or of a generally most involved and abundant complex of cells. We learn that functioning cells make up the tissues and that through combination of the tissues we get the organs whose sum total amounts to the individual being. Also by multiplication of the cells which are a part of the individual, growth and the eventual reconstitution or repair of injuries is effected, and the reproduction of the individual is fostered so that the continuance of the species is in short nothing but a cell function. It is the cell, moreover, which becomes ill and dies causing the death of the whole body of the individual.

SUBCELLULAR AND SUPERCELLULAR ORGANIZATIONS

The validity of the cell theory has been under fire of late. It is indeed possible that certain manifestations of life do exist which we may call *infra* or *subcellular*. Possibly molecular aggregates of sufficient size may form biological units. It is the so-called filtrable *ultra-viruses* and similar chemical and physico-chemical combinations which are the cause of certain extraordinary phenomena and lead us to think along these lines.

The living molecules or combinations thereof making up the micelle have their particular structure, chemical composition, and features which remain constant throughout all the processes of nutrition. The micelle or the molecule, each consists of living matter with vital properties distinguishing it as such. It is further probable that micelles can be divided in the same way as the entity we call the cell, and that in this way they enjoy an independent private life, whence springs their ability to engraft in certain cells which thus become a living culture medium for them. This certainly may be the case with some of the viruses and perhaps also with the genes. The micelle, however, always comes from an organization of molecules and the cell from an organization of micelles.

Let us not forget on the other hand that the organisms come before our eyes as individuals, in the form of a biological supracellular whole.

Polyplastid cells do not lead independent lives but must be subjected to the demands of rigid cohabitation whence they derive certain very important and inevitable vital properties.

None of this, however, can be taken to mean that the functional and morphological element in life is anything other than the cell. The discovery of this truth, that the cell is the essential element, was one of the most outstanding and important steps forward in the progress of biological science and has enabled human knowledge to conceive ideas of incalculable value for the future.

*   *   *   *   *

ROBERT HOOKE 1635–1703

Micrographia, London: Martin and Allestry, 1665.

OBSERV. XVIII. OF THE SCHEMATISME OR TEXTURE OF CORK, AND OF THE CELLS AND PORES OF SOME OTHER SUCH FROTHY BODIES

I took a good clear piece of Cork, and with a Pen-knife sharpened as keen as a Razor, I cut a piece of it off, and thereby left the surface of it exceeding smooth, then examining it very diligently with a *Microscope*, me thought I could perceive it to appear a little porous; but I could not so plainly distinguish them, as to be sure that they were pores, much less what Figure they were of: But judging from the lightness and yielding quality of the Cork, that certainly the texture could not be so curious, but that possibly, if I could use some further diligence, I might find it to be discernible with a *Microscope*, I with the same sharp Pen-knife, cut off from the former smooth surface an exceeding thin piece of it, and placing it on a black object Plate, because it was itself a white body, and casting the light on it with a deep *plano-convex Glass*, I could exceeding plainly perceive it to be all perforated and porous, much like a Honey-comb, but that the pores of it were not regular; yet it was not unlike a Honey-comb in these particulars.

First, in that it had a very little solid substance, in comparison of the empty cavity that was contained between, as does more manifestly appear by the Figure A and B of the XI *Scheme*, for the *Interstitia*, or walls (as I may so call them) or partitions of those pores were near as thin in proportion to their pores, as those thin films of Wax in a Honey-comb (which enclose and constitute the *sexangular cells*) are to theirs.

Next, in that these pores, or cells, were not very deep, but consisted of a great many little Boxes, separated out of one continued long pore, by certain *Diaphragms*, as is visible by the Figure B, which represents a sight of those pores split the long-ways.

I no sooner discern'd these (which were indeed the first *microscopical* pores I ever saw, and perhaps, that were ever seen, for I had not met with any Writer or Person, that had made any mention of them before this) but me thought I had with the discovery of them, presently hinted to me the true and intelligible reason of all the *Phaenomena* of Cork; As,

First, if I enquir'd why it was so exceeding light a body? my *Microscope* could presently inform me that here was the same reason evident that there is found for the lightness of froth, an empty Honey-comb, Wool, a Spunge, a Pumice-stone, or the like; namely, a very small quantity of a solid body, extended into exceeding large dimensions.

THEODOR SCHWANN 1810–82

Mikroskopische Untersuchungen über die Uebereinstimmung in der Struktur und dem Wachsthum der Thiere und Pflanzen (Microscopical Investigations on the Concord of Structure and Growth in Plants and Animals), Berlin: Reimer, 1839.

Though plants outwardly exhibit an immense variety of forms, the simplicity of their internal structure is no less remarkable. Their extraordinary variations in form arise solely from the various types of combination of the simple elementary structures, namely the cells which, though showing modifications to a greater or less degree, are nonetheless inherently similar. The cellular plants consist entirely of cells, some being composed exclusively of single cells, some of groups of cells homogeneously bound together. Similarly the vascular cryptogams in their primitive state consist only of simple cells, whilst the pollen-grain which, according to Schleiden's discoveries, is the basis of the new plant, is in its essence nothing but a cell. In perfectly developed vasculars, the structure becomes somewhat more complex, and up to a comparatively short time ago, their elementary tissues were differentiated into cellular or fibrous, coils or vessels. Investigations into the structure, and particularly into the growth of such tissues, showed that such fibres and coils were, however, nothing but elongated cells, the coiled fibres being merely deposits in spiral form on the internal surface of the cells. Thus the vascular plants consist also of cells, some of which have reached only one step higher in development. The milk-tubes are the only structure not yet broken down into cells, and further research is necessary into the question of their growth. According to Unger, in his *Aphorismen zur Anatomie und Physiologie der Pflanzen* (Principles of Plant Physiology and Anatomy), Vienna, 1838, these also are composed of cells whose dividing walls dissolve into nothing.

Animals present a much greater variation in external form than is to be found in the vegetable kingdom and, particularly in the higher more perfectly developed classes, exhibit also a much more complex structure in their individual tissues. How far exactly does the distinction go between muscle and nerve, between this latter and cellular tissue which by the way is like that of plants only in name, or between flexible and horny tissue, and so on? When we look into the question of development of these tissues, however, it becomes evident that all their manifold forms likewise originate solely from cells, from cells in fact which are absolutely similar to plant-cells and in remarkable accord with them in some of the vital phenomena they produce. The object of the present work is to prove this thesis by a series of observed facts.

First, however, we must give some account of the vital phenomena occurring with plant-cells. Within certain limits, every cell is an individual, independent unit. The vital manifestations of one are repeated wholly or partly in all the others. These individuals are not, however, drawn up side by side to form a simple mixture, but work together in some way unknown to us so as to produce a harmonious whole. The processes which run their course in the plant cells may be reduced to the following main heads—

  1. The production of fresh cells.
  2. The expansion of the existing cells.
  3. Conversion of the contents of the cells and the thickening of cell walls.
  4. The secretion and absorption effected by the cells.

The admirable researches of Schleiden which illuminate this subject so excellently, form the main foundation for my closer observations on these particular phenomena, and I would refer the reader to Schleiden's *Beiträge zur Phytogenesis* (Notes on Plant Development) in *Müller's Archives* for 1838.

Firstly, regarding the production of fresh cells. According to Schleiden, in the phanerogams, except in so far as cambium cells are concerned, this always takes place inside the already ripened cells, and in a really remarkable way from the cell nucleus. In view of the importance of this latter in connection with animal organization, I am including at this point a shortened version of Schleiden's relevant description. The section of an onion, for instance, contains R. Brown's *areola* or cell nucleus, or what Schleiden calls the *cytoblast*, varying from elliptical to circular in outline according to the passage from lenticular to perfectly globular shape of the solid it forms. It is mainly yellowish in colour, though sometimes it turns to an almost silvery white, which because of its transparency is often scarcely to be distinguished. If stained with iodine the shade varies from a pale yellow to very deep brown, whilst its size also varies greatly from 0·0001 in. to 0·0023 in. according to age, to the particular plant and to the particular part of the plant in which it lies. Its internal structure is granular, though the granules forming it are not clearly separated one from the other. Its texture varies from being soft enough to dissolve easily in water to a stiffness capable of undergoing considerable pressure without distortion of shape. Further to these peculiarities of the cytoblast, already published by Brown and Meyen, Schleiden has brought to light a small corpuscle in its interior which in the fully developed cytoblast looks like a thick gasket or thick-walled hollow globule. The appearance seems, however, to differ in different cytoblasts. At times it is possible only to distinguish the outside sharply defined periphery of the ring, with a dark spot at the centre, or more frequently only a sharply defined spot. In some cases this spot is infinitesimal and so minute as not to be in the least bit distinguishable. As it will be necessary often to mention this body in the following discussion, I will call it the *nucleolus* for short. According to Schleiden, sometimes two or even maybe three, or as he has told me personally as many as four such nucleoli may occur in the cytoblast. Their size varies considerably, ranging from the half-diameter of the cytoblast down to practically nothing.

Having done with these preliminary observations, we now pass on to the animal world. The similarity between certain individual animal tissues with those of plants has already been mentioned a number of times. Nevertheless, as is logically correct, no inference has been made from such individual similarities. Every cell is not necessarily analogous in structure to some plant cell, and as to its having a polyhedral shape this necessarily arises from the cells being closely compacted against each other and is obviously no more a token of similarity than the mere chance caused by dense crowding together. We can only draw an analogy between the cells of animal tissues with plant structures of similar elementary identity on the following grounds—

1. By demonstrating that a large part of the animal tissues originates from or consists of cells each having its own particular wall, when it may be possible that such cells do correspond to the cellular elementary structure present in all plants; or,

2. By proving in connection with any particular animal tissue made up of cells, that in addition to its cellular structure, forces similar to those operating in plant cells are at work in its component cells; or, since this cannot be proved directly, by showing that the action of such forces in the shape of nutrition or growth proceeds in them in a way similar or identical to that which occurs in the plant cells.

I was considering the matter from this point of view last summer, when in the course of my experiments on nerve-endings in the tails of frog larvae, I not only saw for myself the beautiful cellular formation of the dorsal cord in these larvae, but also discovered the nuclei in the cells. J. Müller has already proved that the dorsal cord in fish is composed of separate cells provided with definite walls and packed closely together like the pigment in the choroid. The nuclei, so similar in form to the usual flat nuclei of plant cells, might well be mistaken for these, and thus supplied a further point of similarity. Müller had proved, in connection with the cartilage corpuscles discovered by Purkinje and Deutsch, that because of their gradual change into larger cells they must be hollow and therefore *cells* in a wider meaning of the term; Miescher also points out a special class of spongy cartilages cellular in structure. Similarly, nuclei were observable in the cartilage corpuscles. I next succeeded in actually seeing the true wall of these corpuscles, first in the branchial cartilages of frog larva and later also in fish, and the concordance between all of them thus proved that all cartilages possessed a cellular structure in the restricted sense of the word. During the growth of some of the cartilage cells, a thickening of their walls was also perceptible, thus supporting still further the idea of similarity in the process of vegetation in animal and plant cells. Dr. Schleiden at this juncture most opportunely advised me of his admirable investigations on the origin of fresh cells from nuclei inside the parent cell in plants, thus clearing up for me the previously baffling contents of the cells in the branchial cartilages of frog larva, for I now recognized these for infant cells provided with a nucleus. Meckauer and Arnold had already discovered fat vesicles in the cartilage corpuscles. As, soon after this, I succeeded in ascertaining the origin of young cells from nuclei inside the parental cells in the branchial cartilage, the matter was settled. Cells in the animal body showed themselves with a nucleus whose position with regard to the rest of the cell, shape and variants were similar to those of plant cytoblasts, while thickening of the cell wall took place and infant cells were formed within the parent cell from a similar cytoblast, thus proving the development thereof without the aid of vascular connection. This concordance was still further proved by many other details and in this way, so far as concerns such individual tissues, the necessary evidence was obtained to show that these cells did indeed correspond to the elementary cells of plants. I soon surmised that the cellular formation might be a widely embracing, perhaps universal, principle governing the formation of organic substances. Many cells, some with nuclei, were already within our knowledge; for instance, in the ovum, epithelium, blood corpuscles, pigment, and so on. It was an easy step in logic to gather all these known cells into one head and compare the blood corpuscles for instance with the epithelial cells, and consider these—and likewise the cells of cartilages and plants—as tallying with one another and each as facets of the main common principle. This was the more likely, as many points of similarity in development of such cells were already known. C. H. Schultz had already demonstrated the pre-existence of nuclei in the blood corpuscles, the formation of a vesicle around them, and the gradual distension of this vesicle. Henle had observed the gradual expansion of epidermal cells from the lower layers of the epidermis towards the top. The growth of the germinal vesicle observed by Purkinje also served in the beginning as an example of the growth of one cell within another, though afterwards it became more probable that it was not a cell but rather a cell nucleus in its nature thus supplying a demonstration of the fact that not everything of cellular form need necessarily correspond with plant cells. An exact term for those cells corresponding to plant cells ought to be adopted, say either *elementary* cells or *vegetative* cells. On still further investigation,

I continually found this principle of cellular formation coming into its own. The germinal membrane was soon found to be composed entirely of cells, and, not long after that, cell nuclei and later on cells themselves were discovered to be at the origin of all tissues in the animal body, whence all tissues consist of cells or are formed from cells by various means, thus supplying an alternative proof of the analogy between animal and vegetable cells.

RUDOLF ALBERT von KOELLIKER 1817–1905

Mikroskopische Anatomie und Gewebelehre der Menschen (Elements of Normal Histology), Leipzig: Engelmann, 1852.

Our knowledge of the elementary structure of plants and animals is the fruit of work carried out during the last two centuries. This study began with Marcello Malpighi (1628–94) and Anton van Leeuwenhoek (1632–1723), in the period when the microscope, then very simple in design, was first placed in the hands of observers. Neither the Ancients nor mediaeval observers had any real knowledge of the elementary parts which make up organisms, even though Aristotle and Galen did speak of "similar and dissimilar parts," whilst Fallopius had some idea of "tissues" and even tried to classify them in his *Tractatus quinque de partibus similaribus* (Fifth Book of Similar Parts), Vol. 2 of his *Works* published in Frankfort in 1600; nevertheless the intimate structure of these parts remained unknown to such authors.

However brilliant the advances made by science, thanks to the labours of men like Ruysch, Swammerdam and others, such progress had no positive basis on which to establish itself. On one hand, investigators were not yet sufficiently practised in the technique of microscopic observation, and on the other, the cultivation of the less delicate branches, particularly physiology, embryogeny, and comparative anatomy, absorbed too great a proportion of their attention. Thus in spite of certain isolated observations, valuable from many points of view, made by Fontana, Muys, Lieberkühn, Hewson and Prochaska, histology reached no decisive heights during the eighteenth century, and remained at that time nothing more than a hotchpotch of unco-ordinated facts scattered here and there.

Not until 1801 was the genius of one man able to give proper shape to General Anatomy alongside the other branches of anatomical study. This man though certainly he did not enrich histology with any great discoveries did in actual fact bring to fruition something which no one had attempted before him. He classified the material accumulated by his predecessors and considered the physiological and medical application of what was known up to that time and this with so much success that the science of histology can be definitely dated from his work. The *Anatomie Générale* (General Anatomy) by Marie François Xavier Bichat, published in Paris in 1801, is in fact the first scientific work on histology extant and thus represents the beginning of a new era. This great work attained still further importance for the following reason. The tissues are not examined solely from the morphological point of view, but the relationship between them and their functions both in the physiological and in the morbid sense is also discussed.

Considerable impetus has been given to the immense progress made in the field of histology in the present century by continuous improvement of methods of research and of the microscope, and, what is more, the ever-increasing zeal of students. It is not to be wondered at, therefore, that during the past fifty years histology has gone far ahead of what was accomplished in the second half of last century, which may be considered as the initial date of its existence. Over the last thirty years in particular, discoveries have come thick and fast and at the same time it has been possible to link them together in line with their natural logical relationship in such a way that there is now no longer any danger of microscopic anatomy becoming lost in a welter of tiny details, as was formerly the case.

I must particularly point out the year 1838, when Dr. Theodor Schwann proved the unity of constitution of the animal organism, showing that it was originally and uniformly composed of cells and that all the tissues, even the highest morphological structures, were derived from such elements. This is a fruitful theory combining all observations made up to that time, and acquires greater importance as a starting point for further researches.

If Bichat, by the creation of his system, founded the theoretical basis of histology, Schwann with his practical investigations placed it on a foundation of positive facts. Advances in histology, commencing with Schwann, have indeed been of the greatest importance for physiology, medicine and pure science. Certain points glanced at or only briefly mentioned by Schwann afterwards became subjects for more intensive study, for instance, the origin of the cells, the significance of the nucleus,

the development of the higher tissues, the chemical relationships
between cells, and many others.

There still remains much to be done, even after all that has already
been achieved. Without pretending to prophesy, I may say that the
present morphological state of histology will continue until it is no
longer possible to penetrate more deeply into knowledge of organic
structures, when it is no longer possible to discover fresh facts to show
that what we today think of as a simple element is in fact complex. If it
should ever be possible to discover the molecules which make up the
cells, the fibrils of the muscles, the axial fibres of the nerve, etc.; if one
could actually fathom the complete laws of the juxtaposition of parts,
the laws of their development, of growth, and of the functions of these
supposedly constituent parts of the cells; a new era in functional
histology would be on the way. In such an event, the discoverer of the
"law of cell-origin," or of a "molecular theory," for instance, would be
as celebrated, and rightly so, as those who founded the theory of the
cellular composition of all animal tissues, by definitely proving that
"all tissues, both plant and animal, are formed and proceed from the
cell."

RUDOLF VIRCHOW 1821–1902

Die Cellular-Pathologie (Cellular Pathology), Berlin: Hirschwald, 1858.

At the beginning of the present century, Bichat established the principles
of General Anatomy, and new horizons were opened to medical science.
But the progress which histology owes to Schwann has not been greatly
developed nor sufficiently applied to pathology. It is a matter for
surprise to see how little has been achieved in the period since that time.

Each day brings forth fresh discoveries but it also opens up fresh
matters for uncertainty. Is anything positive in histology, we have to ask
ourselves. What are the parts of the body whence commence the vital
actions? Which are the active elements, and which the passive? These
are queries which have given rise to great difficulties, dominating the
field of physiology and pathology and which I have solved by showing
that "the cell constitutes the true organic unit," that it is the ultimate
irreducible form of every living element, and that from it emanate all
the activities of life both in health and in sickness. This manner of

holding life to be a special process may be thrown in my face, and many people may incline to accuse me of some type of biological mysticism, impelling me to set life apart from the great agglomeration of natural occurrences and discount the sovereign laws of chemistry and physics. The object of this course is to show that it is not possible to possess more mechanical ideas than those I myself profess, provided that one seeks to explain what happens in the elementary forms of organism. There is no doubt that the molecular changes occurring inside the cells are referred to some part or other composing them; the final result is however due to the cell from which the vital action started and the living element is not active except in so far as it presents us with a complete whole enjoying its own separate existence.

It has many times been objected that there is no general agreement as to what must be understood by the term "cell." These difficulties date from the work of Schwann himself, who, founding his system on that of Schleiden, interpreted his observations as a botanist in such a way that theories of plant physiology were applied to animal physiology without taking into account that even within the general structure there were plant cells presenting peculiar characters not safely to be identified with the characters exhibited by animal cells. Robert Brown's discoveries contributed greatly to the unification of concepts: I refer to his detection of the nucleus in animal and plant cells. An unimportant role was, however, attributed to the nucleus in the preservation of the cell whilst in contrast its effect in the development of cellular elements was greatly exaggerated.

On the other hand the nucleus, in elements of development, frequently includes an important formation: the nucleolus. According to Schleiden, whose opinion was later adopted by Schwann, the development of the cell takes place in the following manner. The nucleolus becomes the first vestige of tissue, coming to life in the innermost recesses of the formative liquid, blastemes or cytoblasteme, and rapidly acquiring definite volume; small granules separate out from the blasteme and settle around it whilst a membrane is formed to cover these elements. Once the nucleus is completed, a new mass starts to surround it. Some time after, a second membrane starts to form at a definite point on the surface of the nucleus, the well-known "watchglass" form. From this is developed the protoplasm and then the cell membrane.

This theory, admitting the development of the cell at the expense of the free blasteme and requiring the formation of the cell to precede that of the nucleus or cytoblast, is generally known under the name of the "Cell Theory." It would be more precise to call it the theory of the free formation of cells. Nowadays it has been almost entirely abandoned.

In spite of most careful investigations, starting with study of tissues under the microscope, there has never been observed any part which, being ready to grow, or multiply physiologically or pathologically, has not contained nucleate elements as the starting-point of the growth process. In all cases, the first important modifications have their root in the nucleus and this can be confirmed by examining the appearance of the nucleus when an element is about to become transformed. Nothing fresh is created, neither in the shape of complete organisms nor individual elements.

Similarly, the digestive juices do not form the tapeworm, nor do infusoria, algae or cryptogams arise from organic plant or animal waste; in the same way in histology, we deny the possibility of forming a cell at the expense of a non-cellular substance. The existence of a cell presupposes the prior existence of some other cell—*omnis cellula e cellula* (every cell must come from some other cell)—just the same as a plant cannot occur except it be derived from some other plant, or an animal from some other animal. Throughout the range of living beings, plants, animals, or the constituent parts of one or the other, there rules the eternal law of continuous development. Development cannot cease to be continuous, because no particular generation can start a fresh series of developments. We must reduce all tissues to a single simple element, the cell. The whole of an individual made up of distinct tissues is the result of cellular proliferation. There is only one way to form cells, that is by fissiparity; one element is divided after the other. Each new generation proceeds from some preceding generation.

It is not possible to look on an elementary granule, globule, or fibre as the starting point of histological development; indeed there is no reason to suppose that living elements arise from unorganized parts, any more than it can be assumed that certain substances, liquids, plastic elements, matter or blastema give rise to the formation of cells. The formative materials are to be found within the cell, they are neither extracellular nor precellular.

Such are the observations which seem to me to constitute the starting-point for any theory of biology. A single elementary form makes up the living body and remains ever constant. This brings us to a consideration of the higher formations, the plant and the animal, as the sum of the greater or less number of like or dissimilar cells each possessing all the characteristics necessary to life. It is not, however, in a point of superior organization, in man's brain for instance, that the character of vital unity is to be found, but rather in the regular and constant co-ordination of isolated elements. Thus, the individual is the result of a type of social organization, an agglomeration of individual existences dependent one on the other. But this dependence is of such nature that each element has its own special type of activity and even when other parts produce some motivation on it, some impulse or excitation whatever they may happen to be, the function does not emanate the less from the element itself, because of that.

CHAPTER III

STIMULUS AND EXCITATION

IRRITABILITY

GLISSON, a Cambridge don and disciple of Harvey, wrote in 1677: "Every living body possesses the faculty of being stimulated by external influences. That is what we mean by irritability." This idea arises from the psychological concept of sensitivity. Psychological concepts at that time were but rudimentary and purely metaphysical but they were sufficient to embrace the idea that sensation might be the effect of the application of a stimulus or excitation in the corresponding sensorial organ. Thus also was it possible to justify the idea that movement likewise may result from excitation of an organ or muscle.

"Irritability is the property in muscular fibres of perceiving an irritation and reacting thereto. Thus irritation causes contraction and when such irritation ceases the muscle fibres relax. This is identical with natural perception through irritation of sensitive fibres . . . The power of movement is a general property of the organs and inherent to life itself . . . Thus irritability is the prime cause not only of muscular movement but also of life in general. The same thing becomes evident similarly in a state of sickness. Every part of the body suffering discomfort seeks to get rid of it. This effort may be called *irritation*, whilst the parts able to perceive the damage and react to it should be labelled as capable of being irritated."

Almost a century later, Albert Haller (1757) made a distinction between muscular and nervous excitability. "*Incitability*," he wrote, "is a property of the muscular fibres whilst *sensitivity* belongs exclusively to the nervous or muscular system containing nerve fibres. The muscles have the property of contracting quite independently from the nerves, but sensitivity depends definitely on the nervous system." Haller proved by his various experiments that muscle contraction can occur either by exciting it direct or by exciting its motor nerve. The nervous impulse provoked by excitation of the nerve descends through the latter and on reaching the muscle promotes in its turn excitation of the muscle itself and consequent contraction thereof.

"All living matter can be excited," was the axiom laid down by

John Brown in 1788. "Life is excitability and we say living matter is excitable because it can be affected by outside forces." According to whether excitability increases or diminishes, the resultant infirmities are called aesthenic or asthenic.

ELECTRICAL STIMULUS

Luigi Galvani's discovery in 1791 was an important scientific achievement; he found that when two dampened metals, in contact, touch a muscle or motor nerve they cause the muscle to contract. This was a chance discovery, made during other experiments; as Galvani himself explains: "I had dissected a frog and having duly prepared it for quite a different purpose placed it on the top of a table on which stood an electrical machine. The frog was not in contact with the conductor of the machine in question and indeed was a fair distance away from it. One of my assistants chanced to touch the leg nerves of the animal with the point of a scalpel and immediately the lower members contracted as if they had been attacked by violent tetanic spasms. This happened only when the conductor was made to spark." Subsequent to this observation, Galvani found it was sufficient to make a circuit with two different metals closing contact with the animal in order to cause a contraction of the muscles. A few years after this important discovery, Galvani was deprived of his professorship as a reward for his political integrity and died practically in penury.

Baron Humboldt gave the name of "galvanic" to these results, thus honouring their discoverer, and observed that through excitation direct or on a nerve, not only were muscular contractions produced but sensations also were aroused which fomented the working of the glands. This he assumed to be due to what he called "galvanic flux."

Alexander Volta, professor at Pavia, discussed the nature of these observations with Galvani to decide whether they were due to a cause originating in the animal's body or whether this arose from the contact of the two metals reproducing the effect of the electrical apparatus. This latter idea which finally won the day led Volta to the invention of the electrochemical cell which now bears his name. The electric current is produced by the interaction of two different metals or of one metal and carbon—"metals or carbonaceous bodies"—and this is the current which stimulates the organ. This is an important fact, but years later it was found that the converse also is true, that living tissues in action give rise to variations in electric potential.

In 1802, Volta read before the French National Institute in Paris a paper proving the identity of the galvanic fluid with electric flux and also mentioned the physiological properties thereof. The Great Proconsul awarded him a special gold medal bearing the terse inscription "To Volta."

Thus was discovered the agent which on application to the organs causes excitation in them. Nevertheless half a century went by before general use was to be made of electricity as an exciter so easy to handle, so accurate and of such great importance in physiological research.

IRRITABILITY AND ILLNESS

Whilst this was going on, the doctors still indulged in their theoretical speculations on excitation or irritation and its normal manifestations particularly in the pathological field. Rasori in 1807 distinguishes between aesthenic and asthenic illnesses, namely those which he holds to be due to constriction and those which may be due to relaxation. Aesthenia arises from an excess of stimulus and is combated with counter-irritants, such as sedatives, antimony, opium or bleeding. Asthenia is itself a counter-irritant and its remedy is therefore some stimulant such as gamboge, belladonna, strychnine, ipecacuanha or nux vomica.

These ideas were inspired by Francis Joseph Victor Broussais, Breton-born like Laennec, though of vastly inferior calibre to the latter. Broussais was an army doctor in Napoleon's forces, assistant professor at the Military Hospital at Val-de-Grâce, and in 1830 professor at the Faculty of Medicine in Paris. He was a pupil of Bichat and thought himself sufficiently well up to run the Physiological Medicine courses. He lacked, however, the genius of his young teacher.

Broussais was an inveterate theorist who, writing certainly with great eloquence, acquired enormous prestige whilst his works became unaccountably acclaimed, thus causing havoc throughout the world particularly in the nations of Southern Europe and America. His influence extended almost to the end of the nineteenth century.

"Life," he says, "depends on the stimuli setting sensitivity and contractility in motion. Increase in these factors gives rise to irritation which can extend its influence to other parts of the body by means of the *sympathy* exercised through the nervous system. As irritation increases so does the body heat, stimulating the chemical processes of life. Increase in irritation is inflammation, and subsequent increase in heat is fever. The part of the body most sensitive to irritation and most liable to receive and transmit the sympathetic action is the stomach and intestines. Fevers and pyrexia are manifestations of gastro-intestinal inflammation and this being so the basic disorder of the human body is gastro-enteritis. Thus any sickness must be treated with diet and bleeding." Broussais's work was responsible for the fact that all patients for quite a hundred years were atrociously subjected to diet and that the use or rather abuse of bleeding was practised with zest by nearly all doctors, quacks or otherwise. In those cases where bleeding was not thought to be absolutely necessary, the least that was done was to inflict on the sufferer some dozens of leeches. As Castiglioni reminds us, the collection and breeding of leeches was one of the most prosperous industries in Europe thanks to the influence of Broussais.

THE PHYSIOLOGY OF EXCITATION

These outrageous ideas were opposed by the scientific judgment of Claude Bernard, who, by means of his physiological researches, sought repeatedly to bring some order and arrangement into the ideas then current. He published his celebrated book *L'Irritabilité* (Irritability), and in 1878 brought out his *Leçons sur les phénomènes de la vie, communs aux animaux et aux végétaux* (Lectures on reactions common to animals and plants). Irritability is a synonym for excitability. "Every anatomical element, that is to say the protoplasm intrinsic in its make-up, possesses the property of being set in operation and of reacting in a particular way under the influence of outside stimuli." The concept of life is inseparable from that of excitation. "Every living thing is irritable because anything which lives reacts to excitations." There is no vital occurrence which is not the consequence of some type of excitation. "Irritability is an intrinsic factor in perception and motion." Stimuli may be of various types, mechanical, physical or chemical. Claude Bernard was constantly making use of electrical stimulus in his researches, though on occasion he did avail himself of mechanical means when, for instance, he discovered the effects of the glycosuric puncture, whilst in other cases he used stimuli of a chemical nature.

Excitation usually produces some increase in functions. In 1846, however, the brothers Weber had already noticed that electrical stimulus of the vagus or tenth cranial nerve in the neck could cause the heart to fail. They found other instances at later dates. Excitations can therefore be of a stimulating nature or on the other hand they may cause inhibition.

EXCITATION, FUNCTION AND NUTRITION

It was a great step forward in the theory of excitation to connect it with the condition of metabolism, and it. was in Claude Bernard's works that the idea first appeared that excitation is disassimilative and a destroyer of living matter which it thus converts into excremental products. This idea is established pithily in the phrase: "Life is death."

However, up to the time of Hering in 1878, excitation was not identified with disassimilation nor inhibition with assimilation. This concept was crystallized by him in his theory of colour vision. Years later in a famous lecture in 1889, published by the Prague medical periodical, *Lotos*, Hering confirmed and extended his idea of the nature of excitation and inhibition to cover widely different functions. In 1886 Gaskell gave a detailed account of these ideas and suggested the terms *anabolism* and *catabolism* to distinguish the two phases, the assimilative and the disassimilative, of metabolism. Verworn also in his *Allgemeine Physiologie* (General Physiology), which had such a formative influence on physiologists at the beginning of the present century, finally codified the theory with some well-chosen examples.

Every effect of excitation is accompanied and conditioned by acts of disassimilation, of great intensity in the muscle but on the other hand very light in the nerve. These actions are assuredly the two most distinct and characteristic. The effect in the muscle is well known through the evidence of motional and thermal discharges of energy and by the metabolic changes which are inherent to excitation. The nerve effect is a matter of some considerable discussion. Thunberg showed in 1904 that the nerve breathes just like other organs, and Tashiro in 1913 confirmed that gaseous interchange in such nerve is intensified on excitation. An increase in heat production consequently takes place within the nerve, but so negligible that at first physiologists in general thought that no such thing did in fact occur. Later research by A. V. Hill, Gerard, and Zottermann has enabled such functional heat to be measured, on the scale of one millionth of a Centigrade degree for each nervous impulse. One ten-millionth thereof corresponds to the initial heat and the remainder to the heat of recuperation. In 1938, Bronk gave a clear sketch of the development of these important phenomena.

PHYSICAL CHEMISTRY OF EXCITATION

Certain modifications in the physico-chemical structure of an organ occur under excitation alongside the chemical effects mentioned.

Du Bois Reymond in 1848 discovered the negative variation, that is that the field surrounding a part under excitation becomes electrically negative in relation to the remainder of the organ. Hermann in 1867 describes the "demarcation" or critical current, at which any injured portion of the surface or inside an organ is also negative in relation to the uninjured surface. It is easy to infer from these results, that wherever molecular simplification, disorganization, decomposition or catabolism occur, there will be a zone of negative potential as against the unaffected portions. The source of currents due to action, the negative variation which is observed in bipolar conduction, is referable to that of the demarcation currents. Which again confirms that every excitation is accompanied by catabolic discharges.

This entails alterations of the ionic and colloidal balances in those physico-chemical systems represented by the living tissues. Hermann in 1879 assumes that electric charges of one sign or the other, or as we now call them *ions*, are distributed on both sides of a membrane, impermeable or practically so, whence electrical polarization arises. Stimulus increases the permeability of the membrane in such a manner that the ions hitherto held back may pass through it. This is what we know as excitation, as already previously described by Pfeffer in 1873.

Nernst in 1908 propounded a similar excitation theory. The process of excitation would presuppose the concentration of certain ions on one side or the other of the membrane. As Helmholtz stated in 1853: excitation demands a minimum concentration of such ions, otherwise it will not take place.

Both theories agree in considering that the basic condition in the process of excitation is a modification of the respective concentration of ions on either side of the membrane. These theories were modernized in mathematical interpretation by A. V. Hill in 1910 and Lorente de Nó in 1946. In the excitable elements, particularly the nervous elements, there is a state of electric polarization of the cytoplasm in the surface of the cell, whether nerve-cell or some other type, where manifestations develop in the same way as if there were a membrane. Such polarization is tied up with the life of the cell, and when the latter dies, polarization vanishes. There is therefore in every living element a difference between the electric charges residing in it and those of the surrounding medium, which difference is maintained by the living matter forming the surface or membrane of the cell. Excitation is accompanied by instantaneous depolarization whether it be a case of natural excitation, say of a muscle by nervous impulse, or of an artificial excitation, say by electrical stimulus. Such depolarization is arrested in the very act through a reciprocal repolarization process which restores the equilibrium of the electrical charges and brings them back to a state of rest. Obviously such changes must be reversible.

Variations in electric charges have to be pretty rapid to promote excitation before the cell becomes repolarized, and to determine an impluse or activation which is transferred along the cell from the nerve. Yet, on the other hand, they cannot be absolutely instantaneous, since in that case there would be no efficient employment of exceedingly slow changes in electrical potential, that is the electric stimulus, nor of changes too great in speed of the high-frequency currents. The gradient of rise and fall of electric potential and the duration of the change in potential or stimulus constitute, as we have already seen, very important factors in the study of excitation. The physical and physico-chemical changes caused by stimulus on an excitable cell are immediately translated into metabolic effects.

MATERIAL FACTORS IN EXCITATION

From the foregoing we arrive at two important facets of the question. On one hand, there is the possible participation of material factors in excitation, a problem which is still occupying the minds of contemporary physiologists and which at the present time has reached a high stage of development. Langley in 1906 played with the idea of a "receptor substance" intervening when an impulse is transmitted from nerve to muscle; then there was Elliot's "myoneural connection," "intermediate material," and so on.

The latest evidence of the effects of adrenaline in the case of certain organs, of acetylcholine in others, and of the potassium ion and other electrolytes, certain hormones, parahormones and drugs of particular classes is proof that material factors do play a part in the excitation processes.

EXCITATION AND TIME

A second aspect comes from consideration of the time factor conditioning such occurrences. In point of fact, the main object of Nernst's theory was to explain why high-frequency alternating currents did not act as stimuli. J. Müller in 1846 held the view that "the time needed for the transmission of a sensation from the periphery to the brain is infinitesimal and therefore incommensurable." Helmholtz in 1850 showed that this was not necessarily so, that nerve impulses are not transmitted instantaneously, but quite the contrary, they travel at the relatively low speed of slightly less than 100 ft/sec in cold-blood creatures such as the frog, and around three times as fast on the average in warm-bloods such as man, the rapidity of conduction varying with the nature and diameter of the nerve fibres (Gasser and Erlanger, 1927), physiological condition, etc.

This relative slowness in action requires that stimulus, if it is to be effective, shall be applied to the excitable organs with sufficient suddenness (Du Bois Reymond, 1849). This also, however, has a limit. The writer named did not in his researches take into consideration the minimum time the stimulus would have to last, i.e. "the characteristic time" for each tissue as Waller called it.

Hoorweg and Weiss in 1907 showed that if the period of actuation were too short the stimulus had no effect. The work of the Lapicques (1903–9) culminated in the notion of "chronaxy" embodying two parameters, the intensity of stimulus and its period of application in the process of excitation. It was known from previous work that if the stimulus fell below a certain intensity, excitation was not produced, such minimum excitation threshold intensity being denominated by Lapicque as the "rheobase." When therefore it is a question of determining the influence of time in excitation, we have to use a stimulus which is quantitatively sufficient, for instance an electric current of twice the threshold intensity. It is then necessary to measure the minimum effective time, by shortening the period of stimulus gradually, until we reach the minimum or "chronaxy." Obviously a body will be the more excitable the lower its chronactic number.

Modern researches have shown how excitation develops in relation to time, such periods being reckoned in thousandths of a second. In the first place, excitation does not commence simultaneously with the application of the stimulus, but only after a period of latency which usually is fairly short. Einthoven thought that there was no such latent period, and that the delay observable in the response of a muscle to direct excitation or to a nerve impulse depended upon purely mechanical conditions such as the inertia of the system or the elasticity of the organ or some such thing. Exact measurements by Cooper and Eccles in 1930 showed that in a muscle the actual latency period is one-thousandth of a second; similar values were observed for nerves. In muscles, impulses travel slightly more slowly than in the nerves.

The excitation process continues splitting into various phases according to the metabolic changes taking place. Some considerable time after Marey had studied the phase of inexcitability in the heart,

Gotch and Burch in 1899 found that there are also moments of refractory condition in the nerve; first an absolute period of total inexcitability, then a relative period of progressive recuperation from the excitability. Adrian and Keith Lucas (1912–21) discovered that after the refractory condition there developed a period of supernormal excitability followed finally by normal repose.

It is not difficult to approximate these facts to respective metabolic conditions. Inexcitability in the course of full excitation must be attributed to the disassimilative flow produced in those particular instants. Supernormal excitability is connected with the assimilative reaction which balances the preceding functional catabolic phase. Researches by the aforesaid Lucases in 1912, by Lillie in 1922 and of other still more recent investigators confirm this.

SIGNIFICANCE OF EXCITATION IN LIFE

We are gradually getting to know more exactly what excitation may really be, but the process still possesses many unknown features and presents itself in many guises affecting all forms of animate matter. In life there appears a certain functional tonality, a primary biological feature or property allied with a particular metabolic tone. Tonality can be modified in either direction: increase signifies excitation, and decrease, inhibition, producing corresponding modifications of the metabolic tone, namely dominance of catabolism or of anabolism.

Such variations occur in a wavy ever-moving line resulting from the influence of stimuli which previously had been thought to be always outside the animate being, but which as we now know may also be internal to it. This is why the life "quantum" is not uniform but varies constantly in accordance with external and internal conditions. This possibility of suitably exaggerating or diminishing the various functions, a possibility which may be applicable to all animate manifestations, allows the adaptation of living bodies—animal principally—to changing conditions to which they are ever subject, and is a prime, positive and essential condition for the maintenance of existence.

*　*　*　*　*

GIOVANNI ALFONSO BORELLI 1608–79

De Motu Animalium (Motion in Animal Organisms), Rome: Bernabò, 1680.

The nerve fibres are not solid, hard-packed and impermeable bodies, neither are they hollow empty tubes like a reed, but they are conductors made up of a spongy matter similar in structure to elder pith. Thus the marrow of fibre may be easily moistened by volatile juice from the brain with which such fibres are in communication and can moreover be saturated until they swell up like a sponge soaked in water.

All writers now admit that the nervous juice is not an accumulation of wind or of inhaled air but a liquid of the consistency of spirits of wine. This nervous juice even though spirituous and active is always of a corporeal nature and is unable without physical contact to intensify or lower the animal spirits. It is by means of its corporeal presence that on mixing with the animal spirits which also are of a corporeal nature it heightens the action of such spirits or drives them out or changes them in some way. It is impossible to imagine that nervous action may be developed without some local movement of the nervous juice, passing along the nerve in one direction or the other, either from the surface to the brain generating sensations, or from the brain to the muscles thus causing movement. In this latter direction, the nervous humour as conceived by Willis fulfils a double function. It not only gives rise to visible movements through use of the muscles but also exerts a variable power of nutrition whereby the foodstuffs supplied through the blood are converted into living flesh.

If one of the ends of a nerve is compressed, nipped, struck or punctured, the disturbance in the form of an impulse or wave is communicated to the opposite end, since in proportion to the nearness of the parts arranged in ordered series the impulse keeps flowing successively throughout the whole nerve.

From this it follows that the fibres or spongy ducts of certain nerves distending with spirituous liquor can be vibrated through the gentle motion of the spirits by means of which the brain communicates to the outer reaches the actions impelled by the mind. The impulse will descend through the whole length of the nerve and then by irritation of the end there will be driven out or discharged from that extremity a few droplets of spirit into the corresponding muscle causing a type of explosion or ebullition which makes the muscle contract and become tense.

Again, if one extreme of a sensory nerve terminating in the skin, nose, ears or eyes is compressed, pinched or tickled the spirituous juice of such nerve transmits the wave or impulse to the spot in the brain where the nerve ends. And here, the power of the sensitive pith in relation to the region of the brain affected together with the force of the disturbance and the manner or quality of the nervous movement, will produce an assessment of the object causing such movement.

FRANCIS GLISSON 1597–1677

De Ventriculo et Intestinis (The Stomach and Guts), London: Brome, 1677.

Now I come to the second reason which shows that animal spirits are not the immediate cause of the feelings and movement. It is the exception in nature for the same conduit to carry matter in both directions, flow out and flow in. In breathing, the air, inhaled as well as exhaled, certainly travels through the same organs, but the cartilaginous pipes forming the breathing passages are made for that special purpose and the thorax expands and contracts alternately. Also in the viscera peristaltic and antiperistaltic motions take place but these are necessary for the preparation and conveyance of the chyle. In the case of the nerves, it would be difficult to conjecture why the spirits flow sometimes forward and sometimes backward. The sensation produced by a hard object lasts for many hours in certain cases. How can we imagine that all this time the animal spirits might be going to and fro from the sense-organ to the nerve-centres and that the sensation should remain continuously throughout such time? This problem is beyond me. It is said, moreover, that these spirits cause the fibres to swell in motion. Certainly the muscles contract in length but they also thicken at the same time. The experiment I shall describe below contradicts the idea that muscular contraction is something of the nature of an explosion or outburst, that is to say a distension or swelling of the muscle.

Take an oblong glass tube of suitable size and shape, and to its mouth affix another upright tube ending in a funnel. Get a strong muscular man to thrust his bare arm into the large tube and stop the mouth of the tube immediately with clay around the start of the arm near to the shoulder, in such a way that there will be no escape of the water with which the space between the walls of the tube and the arm is to be completely filled by means of the funnel at the top end of the vertical tube.

The water must reach a certain level in the vertical tube. Instruct the patient first to contract his arm muscles strongly by clenching his fist and then to relax them right away. It will be seen that during contraction the level of the water falls and that it rises again on relaxation. Clearly, therefore, during contraction the muscles do not distend but quite the contrary they decrease in volume. This shows that the fibres contract by reason of their intrinsic vital activity and not because they stretch and consequently become shorter from a flow of vital or animal spirits bearing orders given out by the brain and causing such voluntary movements.

This intrinsic property of the muscle is similar to that of the gall bladder which though constantly receiving bile only empties itself at particular instants. This is the property of being irritable and such property of certain organs may be called "irritability."

WILLIAM CULLEN 1710–90

First Lines of the Practice of Physic, London: Murray, 1774.

While the doctrines of Stahl were prevailing in the University of Halle, Dr. Hoffmann, a professor in the same university, proposed a system that was very different. He received into his system a great deal of the mechanical, Cartesian, and chemical doctrines of the systems which had appeared before. But, with respect to these, it is of no consequence to observe in what manner he modified the doctrines of his predecessors, as his improvements in these respects were no ways considerable, and no part of them now remain; and the real value of his works, beyond what I am just now going to mention, rests entirely on the many facts they contain. The merit of Dr. Hoffmann and of his works is that he made, or rather suggested, an addition to the system, which highly deserves our attention. Of this I cannot give a clearer account than by giving it in the author's own words. In his *Medicina Rationalis Systematica*, Volume III, Part 1, Chapter 4, in the forty-seventh and last paragraph of this chapter he sums up his doctrine in the following words—

"Defects in the microcosmic movements of solids may explain, through nervous affections, a certain type of internal malady. Alterations which run their course through the nerves or nerve membranes will give rise to anomalies outside the usual course of nature, sometimes serious, sometimes of little consequence. Careful observations confirm the reality of unhealthy motions which have their origin in the alteration of influences emanating from the nervous organs, which by way of the nerves, extend their action to all parts of the body. It is these conductions of the subtle juices which, alternating in condition from systaltic to diastaltic, arrive in the gut, belly, gullet and anus, in the arteries and bile, salivary and urinary ducts, muscles and subcutaneous regions and the very membranes of the brain and medulla, in the sensory organs and ligaments which enfold the bones and form the joints. There is no pain, inflammation, spasm, loss of power, fever, alteration of the humours or of the secretions in which such influences do not play some part."

It is true that Dr. Willis had laid a foundation for this doctrine, in his *Pathologia Cerebri et Nervorum* (Pathology of the Brain and Nerves); and Baglivi had proposed a system of this kind in his *Specimen de fibra motrici et morbosa* (Examination of Morbose and Motor Fibres). But, in these writers, it was either not extensively applied to diseases, or was still so involved in many physiological errors, that they had attracted little attention; and Dr. Hoffmann was the first who gave any tolerably simple and clear system on the subject, or pointed out any extensive application of it to the explanation of diseases.

There can be no sort of doubt that the phenomena of the animal economy in health and in sickness, can only be explained by considering the state and affections of the primary moving powers in it. It is to me surprising that physicians were so long of perceiving this, and I think we are therefore particularly indebted to Dr. Hoffmann for putting us into the proper train of investigation; and it every day appears that Physicians perceive the necessity of entering more and more into this inquiry. It was this, I think, which engaged Dr. Kaaw Boerhaave to publish his work entitled *Impetum faciens*; as well as Dr. Gaubius to give the Pathology of the *Solidum vivum*. Even the Baron van Swieten has upon the same view thought it necessary, in at least one particular, to make a very considerable change in the doctrine of his master, as he has done in his Commentary upon the 755th Aphorism. Dr. Haller has advanced this part of science very much by his experiments on irritability and sensibility. In these and in many other instances, particularly in the writings of Mr. Barthez of Montpellier, of some progress in the study of the affections of the Nervous System, we must perceive how much we are indebted to Dr. Hoffmann for his so properly beginning it.

ALBRECHT VON HALLER 1708-77

Elementa Physiologiae Corporis Humani (Physiology of the Human Body), Lausanne: Bousquet, 1757.

Not only in the animal kingdom but also among the plants there is to be found a widely spread contractile force which causes the fibres to approach one another. This appears to us to be the cause of cohesion in general, which is evidenced when a fibre stretches or tightens and then relaxes returning to its original length, the cause being an elastic force of both a physical and a cohesive nature. On the other hand a live or dead fibre is also given to shrinking when heat is applied or if it be submitted to some other treatment. Such behaviour is to be seen in all living bodies which are not of too soft a nature, such as the cerebral pulp, or too hard, like the bones or teeth.

Alongside this, however, we must consider a special type of contractile force inherent to the muscles. In a living animal or in one which has just died there frequently occurs in the muscular tissue a rapid and striking contraction, during which the end of the muscle is brought nearer to its middle and after relaxation pushed away therefrom. This contractive motion cannot normally be produced spontaneously except when brought about by specially applied irritations in the live animal. The force causing contraction in the live state cannot help but be something different from that provoking retraction in the dead body, since each of these processes differs in the laws which govern it, both as to duration and as to the point at which it develops. The force at play in the live being may be called "inherent force" (*vis insita*) and the tissues possessing it may be called, as Glisson suggests, irritable. This inherent force, quite different from elasticity and from that causing retraction in dead tissues, both of which latter are common to all the fibres, is characteristic of the muscular fibre and constitutes a particular feature thereof distinguishing the physiological character of such fibre, which is rendered distinct from the very fact of its being irritable. We are dealing therefore with a special force different from any other, which may be classified among the causes of motion, the ultimate source whereof is unknown. It is inherent in the fibre itself and does not reach it from any other part remote therefrom.

Therefore the heart and the intestine containing the muscular fibre are by their very nature irritable; they are able to function by themselves alone without any influence reaching them from outside. Some writers are inclined to call this the "vital force," but such term does not seem justified in my view, since it is possible to demonstrate effects caused by it even after life has departed. I prefer to call this the "inherent or intrinsic force" of the muscle. We have seen that it comes into action through the application of various stimuli, but the very interesting fact also comes out that a very small stimulus may provoke a movement of great intensity.

Alongside this force there is another which may be exerted on the muscle. It is carried from the brain through the nerves and sets the muscles in motion. This we call the "nervous force" (*vis nervosa*). Like the inherent force it survives the death of the body, as may be demonstrated in cold-blooded animals recently dead, wherein without either sensations or voluntary movements, contractions are produced to irritate the nerves ending in the muscles, provided such muscles remain damp and intact. The same may also occur in warm-blood creatures.

The ancients attributed to the nerves a subtle and refined "humour" of great fluidity. The word "humour" however suggests the idea of something indolent, lazy, sluggish, slow, dawdling or heavy, for which reason we prefer to call the active agent "spirit," that is, something movable, invisible and powerful like the air. These theories persisted for many centuries. At last, like so many others which were good enough for antiquity, they threatened their own destruction. Thus it is that an important school like Stahl's came to think that the soul might act directly on the whole body without needing any intermediate instrument to transmit orders from a distance. In point of fact, the existence of certain spirits, which would be nothing more or less than material principles, is somewhat doubtful.

It has been put forward, as proof of the non-existence of such a class of spirits in the nerve, that the latter does not stretch or swell above a constriction which would not allow such spirits to pass on to their fate. This objection is invalid, since the fluid might be so tenuous that the constriction would be no obstacle to its transmission. There are no proofs, on the other hand, that these hypothetical vital spirits are of an alcoholic, acid, sulphurous, albuminoid or combustible nature. Other writers have assumed that they were aerial in nature and during the present century the name "ether" has become fashionable to such an extent that it is customary to attribute to the ether everything whose causes not being susceptible of practical demonstration remain unknown, such as light, the force of gravity, and magnetism among others. In line with this tendency certain people assume that the spirits, the "nervous fluid," are of an ethereal nature and compounded of ether. This idea does not seem justified to me, nor that such agents should be credited with electrical properties.

Of what nature, then, can these spirits be? A spirit must be an element with its own essential properties distinct from everything else, of so tenuous a nature that it cannot be perceived by any of our senses, but coarser than fire, ether, electric or magnetic flux, since it can be contained in ducts, restricted by binding and produced and maintained by food. All of which allows us to assume, since light is different from fire and the material of a magnet differs from one to another, whilst air and ether are two entirely different things, that what we call spirits, the inherent force as well as the nervous force, may be some special element peculiar to living organs and that we may know them only by their effects! A property of living matter.

CLAUDE BERNARD 1813–78

Leçons sur les Phénomènes de la Vie, communs aux Animaux et aux Végétaux (Lectures on Reactions common to Animals and Plants), Paris: Baillière, 1878.

Protoplasm, the agent of the phenomena of organic creation, not only possesses the power of chemical synthesis which we have previously examined, but, furthermore, exhibits for the purpose of setting such power in play, the faculty of being "irritable," and consequently through its irritability the power of motion. It can, in fact, react by contracting under the action of external excitants.

Happenings in life are not spontaneous manifestations of any internal vital principle. They are on the contrary the result of conflict between living matter and external conditions. Life is manifested through the reciprocal connection between these two factors in the phenomena of sensitivity and movement which are considered in general to be of a high order, as well as by those due to physico-chemical action.

This continuous relation between organic substance and its surrounding medium is a general feature of organic life, just as it is of the so-called animal life. Nutrition, like sensitivity or movement, shows in more or less complex form this property of response to excitation. This faculty, which is an essential condition of all life phenomena in plants and animals, exists in its simplest form in the protoplasm. It is "irritability."

In generic terms we may define irritability as "that property possessed by every anatomical element, that is to say protoplasm, of being set in motion, and of reacting in a particular way under the influence of excitants."

Every vital manifestation demanding the combination of set conditions or external excitants is a manifestation of "irritability." Sensitivity or perception which is in the highest degree a complex phenomenon cannot be other than a particular case of irritability, which is the only fundamental biological property common to both living kingdoms.

We must consider irritability a simple form of perception and, reciprocally, in perception we find a very high form of irritability, that is of the property common to all live beings of reacting according to their particular nature under alien stimuli.

Linnaeus made perception the distinguishing feature of the animal kingdom, saying: "Plants exist, animals feel" (*Vegetalia vivunt, animalia sentiunt*).

If, however, we examine this sensibility, seeking to acknowledge it as a higher attribute of animal life, we soon become convinced that it is not a simple "property" but a complex vital manifestation corresponding to a "function." Moreover it is necessary to distinguish between the functions of a live body and the properties of the organic matter composing it. Perception may be a complex phenomenon inherent in certain organisms, but it arises from a simpler general source.

Broussais did not accept any essential property of organic substance other than irritability, which implied consequently perception, contractility and other secondary actions. Virchow likewise held that vital phenomena possess the essential feature of irritability, a generic term comprising all other vital properties.

This theory is to be found in its incipient state even in Bichat, who used the word "sensibility" or "perception" under all circumstances, thus causing considerable confusion; nevertheless what he understood by sensibility was what we now call irritability. In his time no distinction had yet been made between the two concepts. Bichat recognizes in animals an "animal sensibility" and on the other hand a "vegetative" or "unconscious sensibility" residing in the vegetative organs and translatable into visible actions on the part of the organs when they are aroused by an internal stimulus. It may, however, happen that such reaction to physiological or artificial excitants is not translated into any visible sign or movement, and yet nevertheless the reaction does develop, being lost intrinsically in the nutritional motions which do not come into evidence except by their subsequent effects. This is what happens particularly in the case of plants, wherefore Bichat attributed to plants and certain parts of animals an "imperceptible sensibility," that is, something not evidenced by any visible sign whatsoever.

Whatever opinion one may hold regarding the choice of these terms—conscious perception, unconscious, unknowing, imperceptive or senseless sensitivity—it must be recognized that they stand for actual facts and tally with a proper feeling of reality. All actions of the organism are acts provoked by external or internal, physiological, normal, abnormal or artificial stimuli. For their fulfilment they require perception; if this word is not to cover anything other than the power of reacting to stimulus, let us call it "irritability." It is of course correct that in such reaction there may be various stages, from "purely nutritional or trophic reaction"—imperceptible—up to "motor reaction" which is clearly appreciable, and finally the highest stage of "conscious reaction."

CHAPTER IV

BIOCATALYSTS

CATALYSIS

IN 1812 Kirchhof discovered that if starch were heated with sulphuric or other strong acid solution it was converted into sugar without any change or loss in the acid itself. Thénard noticed that hydrogen peroxide solution decomposed, giving off oxygen in the presence of undissolved bodies, such as finely divided platinum, silver, manganese dioxide or blood fibrin, not entering into the reaction, whose mere presence was sufficient to produce this effect. Sir Humphry Davy showed that platinum in the powdered or spongy state, and therefore of greatly increased surface area, can oxidize alcohol vapour exposed to the air without the application of heat.

Berzelius in 1837 called general attention to the meaning of these facts: "In reactions of this nature there is present a chemical force quite distinct from any hitherto known," and he proposed to call such operations *catalytic* in view of their being absolutely different from *analytic* actions resulting from the use of those reagents normally used in chemical research. In ordinary analytical processes the reagents come in according to the laws of affinity of normal chemical relationships, whereby they are transformed whenever they are present. On the other hand, in catalysis, the factors giving rise to the action have their effect merely by being present and in themselves suffer no intrinsic change whatever.

Prior to this basic statement by Berzelius, which is one of the cornerstones in the history of chemical science, Dubrunfaut in 1830 had already made one important discovery, namely that fermented barley extract or malt converts starch into glucose even in the cold state, similarly to the way an acid acts in the presence of heat as above mentioned. Shortly thereafter, Payen and Persoz in 1832 succeeded in separating the amylolytic principle from malt by extraction in water and precipitation with alcohol, giving it the name *diastase*. They considered this agent identical with the natural *ferments*, which cause souring in wine or milk. Robiquet consequently showed that bitter almonds contain a protein substance which breaks amygdalin up into prussic acid and glucose; this active protein was christened *emulsin* by Liebig and Wöhler.

The fermentative properties of the gastric and pancreatic juices came under consideration practically at the same time. Pepsin was discovered by Eberlé and Schwann, and trypsin by Corvisart. These newfound substances were to be of the greatest value in bringing the study of digestive chemistry nearer to that of the catalysts. Thus it is that the digestive diastases at that time came to be the most typical examples of physiological catalysis, and that is how diastasis and catalysis acquired their greater renown as factors in the vital processes. The idea got abroad that certain animal organisms contained or produced diastases in just the same manner as did certain vegetable tissues.

FERMENTATIONS

Pasteur's researches on the processes of fermentation extended from 1857 until 1870. Some score years previously, in 1838, Cagniard de La Tour asserted that fermentation was due to microscopic beings. Berzelius opposed this idea that fermentation should be one of the vital processes. "The case of fermentation," he wrote, "is similar to that of the decomposition of hydrogen peroxide in the presence of platinum, silver or fibrin. A catalytic action like any other."

Pasteur's great work definitely proved the effect of various yeasts and bacteria in different fermentation processes and determined the conditions in which such processes could develop. "Fermentation is the work of living organisms," he wrote, and very soon proved the fact that not only are single-celled micro-organisms possible factors of fermentative change, but also that cells from metazoic or metaphytic tissues existing communally in polyplastid organisms might also play their part in the process. Fermentation might therefore be shown to be a vital property inseparable from the structure and functions of the cell.

Liebig, from 1839 onwards, maintained that ferment is a soluble matter of protein nature working through catalytic action, and that structural or functional life was not essential to demonstrate its effects and properties. Ferment, he said, was neither an organized nor an organic product with all the complexities which organized life would entail. Liebig's "soluble ferment" was the same as Payen and Persoz's "diastase"—later on Duclaux was to propose for general acceptance the generic use of the word *diastase* to distinguish soluble ferments—it is the same thing as Kühne's "enzyme" or "zymase," etc.

DIASTASES AND PHYSIOLOGY

The labours of Traube (1838), Berthelot (1860), and Hoppe-Seyler (1876) finally facilitated the correlation of these various theories, whether life or mere catalysts, were responsible for fermentation. Numerous diastases were discovered formed from the most diverse types of cell. That is how cells make diastases and how the fermentative effects of such cells become evident through the diastases. Life is, therefore, catalytic through the action of its diastases originating from vital actions and comes about at the same time through catalysis; which suggestion does not, however, prevent the parallel existence of catalysts not endowed with life and even of an inorganic nature.

These concepts found definite confirmation as a result of the researches made by Buchner and published in 1897; yeast extracts totally free from cellular matter even in fragmentary form, containing therefore only dissolved matter formed from living cells and separated entirely from cell plasm, were capable of developing alcoholic fermentation if they came into contact with glucose. The diastases of alcoholic fermentation are manufactured by yeast cells, that is living organisms, in the same way as pepsin, trypsinogen, amylase or lipase are manufactured by the gland cells of the stomach or pancreas. Diastases are therefore merely products of metabolic action or cell functioning.

These ideas became still more positive when the intervention of the diastases was speedily discovered to be a factor in all nutritional processes and an influence in the development of all functions. Diastases became accordingly something more than mere digestive factors. Wittich in 1873 proved the diastatic nature of the evolution of glycogen in glucose in the liver. He isolated the diastase by solution in water and glycerine and precipitation with alcohol. Claude Bernard in 1877 after having many doubts on the matter, confirmed Chauveau's assumption of 1856 that glucose was consumed by the tissues. Similar consumption occurs in the blood, which when protected from putrefaction at room temperature in the laboratory rapidly loses its sugar content. Lepine in 1890 and Arthus five years later attributed the glycolytic diastase or diastases present in the blood to be of leucocytic origin, but immediately pointed out also that the tissues in general and the muscle in particular produced similar diastases in greater quantity.

Starting from this point, discoveries of numerous and distinct diastases from all the tissues came thick and fast. Each act of metabolism arises from the action of the corresponding diastase or diastatic group, and as time goes on we are daily increasing our knowledge of fresh details in regard to the intimate relationship between nutritional reconversion and the functioning of the organs. The study of diastases thus covers a very wide field, particularly during recent times. As Fearon said in 1940: "Since 1900, research on the diastases has reached such a stage of development that each single diastase or a single feature of a particular class of diastase is sufficient to occupy fully an entire group of research workers."

THE NATURE OF CATALYSIS

Ostwald, in 1896, defined catalysis as an acceleration of the corresponding chemical reaction without modification of the state of balance of such reaction and without affecting therefore the amount of energy liberated or absorbed thereby. From that time on, researches have been repeatedly started for the purpose of investigating the nature and course of diastatic actions. Among the many scientists concerned, we may mention Jacobus Henricus van 't Hoff, Arrhenius, Armstrong, Duclaux, Sabatier, Födor and others who busied themselves with the question of diastatic catalysis and conditions leading thereto. They were able to assert once more that diastasis is one particular case of catalysis.

According to Hughes (1933) a diastase is any substance which by its mere presence allows certain molecules to undergo chemical change through the effect of a critical quantity of energy less than what would be necessary were the catalyst not present.

There are many types of catalysts at work in living matter; sometimes they are the hydrogen, hydroxyl or iodine ions and their like, at others the catalyser may be a dissolved organic particle, metabolite, hormone or vitamin, an inorganic colloid like the metallic suspensoids or lyophobic colloids, distinct organic particles, or diastases of some special type; and perhaps, finally, combinations energetic in action rather than of a specific material effect, for instance, porous walls (Becquerel, Giraud), radiations (Victor Henri, Achalme), or membranes or other cellular structures (Bayliss 1919, Willstätter 1933).

THE BIOLOGICAL SIGNIFICANCE OF CATALYSIS

It is through catalysis that vital phenomena are developed in living organisms, since only by this means can one explain how otherwise impossible chemical reactions take place. There is no single physiological process which is not under the effect of many vital catalysts.

Diastases, hormones, parahormones, harmosones, vitamins and certain ions govern the activity of the organism as a whole as well as each of its parts, and control the appearance and development of the most diverse phenomena of life.

The vital catalysts are countless in number and act in a systematic manner co-ordinated rigorously within the framework of the being, their intrinsic potentiality being most intense. Haldane for instance has estimated that a molecule of catalase can decompose somewhere about 200,000 hydrogen peroxide molecules per second. Conversely, each molecule of the substrate will come into contact with a large number of diastase molecules and thus be converted. Haldane's calculations show that the average atom undergoing oxidation in the course of its metabolic life through the cell encounters no less than 100 catalysing molecules one after the other. These facts show that the immense metabolic activity present in the living state can be explained by the presence and working of suitable catalysts. It is not possible to guarantee that these figures completely represent the true state of affairs, but, even though they be mere approximations, the extent and intensity of the biocatalytic function can thus be inferred. Such catalysts must unfailingly be present in all living bodies, whether plant or animal, and it is because of them that life continues to exist. Without catalysts there would be no metabolism, and without metabolism there would be neither function nor morphogenesis. It is impossible to conceive of any vital manifestation of any type without the assistance of the relative catalysts.

CATALYSTS AND LIVING CONDITIONS

Catalysts enter into the appearance and development of the vital manifestations. Conditions most favourable to them occur naturally in circumstances which are likewise best for the maintenance of life: moisture, for instance, since biological catalysts must be dissolved in water—the substance cannot act unless dissolved; warmth, since they act best and with greater intensity at the body temperature of warm-blooded creatures; particular osmotic pressure and a predetermined acid-base balance individual to each one of the various groups of diastases; ionic surroundings and the presence of certain substances also help.

In this way complex diastases may be observed, and the functions of co-diastases, activating or inhibitive agents, and antidiastases must also be borne in mind. Diastases may have their source in the inactive form of prediastases and then be set in motion by various means.

Other vital catalysts, like the hormones or vitamins, demand the help of particular circumstances, or contrariwise find their physiological effects opposed by the presence of particular types of disturbing factors. Various metabolites, of which we could mention many, such as glutathione, diphenyloxylic acids, adenylic acid, flavin, etc., exert powers of fermentation analogous to those of the diastases, probably without any diastase whatever having to intervene. As may be seen, the catalysing agents in organic life are of a varied character, the wealth of their activity belongs to life itself and the need thereof is equally intrinsic to requirements for the continuance of life.

In little more than a single century immense progress has been made in theory, particularly as far as concerns biological catalysis and the real importance of those phenomena, originally thought to be nothing more than individual simple applications of organic chemistry of no greater interest than, say, their value in industrial application; the fermentation processes, for instance, have been effectively developed. It is obvious that a lot more still remains to be done; catalyst mechanisms have been shown to be indivisible and of the same essence as all the processes of life. "Life is ferment," as Paracelsus wrote in the sixteenth century.

* * * * *

JÖNS JACOB BERZELIUS 1779-1848

Lärebok i Kemien (Animal, Plant and Mineral Chemistry), Stockholm: Nordström, 1808-28.

A Force Different from any Hitherto Known

It is indeed surprising that the blood-stream supplying all organs alike should, without the help of any other liquid whatsoever, produce saliva, milk, urine, and so forth. In 1812 Kirchhof made a discovery which gave us a first glimpse at the life processes though remaining still far from a full explanation thereof. Kirchhof observed that starch could be converted into glucose by the action of dilute sulphuric acid, the latter remaining unchanged and unconsumed in the process. Thénard showed subsequently that hydrogen peroxide can be decomposed by finely divided platinum and other substances in suspension. Platinum itself in the spongy form is capable of oxidizing alcohol at room temperature.

It is common knowledge that the conversion, for instance, of sugar into carbon dioxide and alcohol takes place during fermentation under the action of an insoluble body which we call *ferment*, and that animal fibrin, coagulated albumin obtained from various plants, cheese and similar substances behave like such ferment though with less intensive effect. Such reactions cannot be explained by the laws of chemistry, as a double decomposition taking place between the ferment and the sugar or any similar reversible reaction. It may perhaps be better to compare such conversions with inorganic processes such as the well-known decomposition of hydrogen peroxide in the presence of platinum, silver, fibrin, and so on, from which we must assume that the action of the ferment may be of similar character.

I do not by any means imagine that we are dealing with chemical forces different from those exhibiting themselves as properties of matter. In some cases it is quite possible to explain reactions of the type under consideration by availing ourselves of the well-known laws of chemistry and physics. Therefore I propose calling the power of substances causing the conversions above mentioned under the conditions indicated *catalytic*, contrasting the word *catalysis* with the term *analysis*, so much in vogue for signifying the separation of the components of a body by means of chemical affinity in the normal sense. Catalytic power seems to consist in the faculty certain substances exhibit for setting in action potential affinities and chemical activities which may be lying dormant at a particular temperature and are presumably something quite different from the real affinity between the bodies, since their action occurs through their mere presence.

There are well-justified reasons to suppose that in living animals and plants, thousands of catalytic processes take place between the organic fluids and tissues, whence we get the formation of countless heterogeneous chemical compounds, starting with plant saps or animal blood whose composition does not change through the fact that so many and various products are manufactured. It is likely that some time in the future we shall find out the cause of it all is the catalytic power of the tissues forming the organs of the living body.

Every solid living tissue placed in contact with a liquid seems capable of acting chemically upon the substance in solution, whilst at the same time the dissolved matter can react without any interchange of elements ... Catalysis is both chemical and physical in its nature.

JUSTUS VON LIEBIG 1803–73

Die Thierchemie oder die organische Chemie in ihrer Anwendung auf Physiologie und Pathologie (Organic Chemistry in its Application to Physiology and Pathology), Brunswick: Vieweg, 1842.

Attention has recently been called to the fact that a body may exert influence by mere contact upon some other body which is thereby decomposed or combined. Platinum, for instance, does not decompose nitric acid neither at the boiling point of the acid nor in the finely divided non-reflecting state (spongy platinum), but an alloy of silver and platinum is easily dissolved in nitric acid. The oxidation suffered by silver gives rise to a similar change in platinum. In other words, platinum in the presence of silver undergoing oxidation, acquires the property of decomposing nitric acid. Copper does not decompose water even when it is treated with boiling dilute sulphuric acid, but an alloy of copper, zinc and nickel dissolves under such conditions, hydrogen gas being liberated.

These and similar reactions cannot be attributed to "special forces" differing from those governing chemical affinities, forces which would be set in motion by a touch as light as a feather. The unstable substances which are so easily decomposed give the greatest number of instances of this type of phenomenon. The components of these bodies are found in such a state of tension that the slightest modification of their atomic balance overcomes the effects of chemical affinity. They exist only by the force of inertia and the slightest shock or disturbance is sufficient to destroy the mutual attraction of their components.

Hydrogen peroxide belongs to this class of substance; it is decomposed by any product capable of attracting the oxygen which goes into its composition and even by the mere presence of certain substances such as finely divided platinum or silver, which themselves do not enter into combination with any of its constituents. The cause of the sudden separation of the elements in hydrogen peroxide has been held to be quite distinct from that giving rise to a common chemical decomposition, and has been called the "catalytic force." It should be remembered, however, that the presence of platinum or silver has no other effect than to accelerate the decomposition of the hydrogen peroxide, which would take place even without the intervention of those metals but at a very much slower rate.

Certain organic bodies seem to enter spontaneously into a state of fermentation or putrefaction, especially those containing nitrogen. It is worthy of note that very small quantities of such substances in that state possess the power of bringing about similar transformations in unlimited quantities of substrate. Thus a small sample of grape-must in fermentation added to a large amount of the same liquid in the still state causes the whole lot to ferment. Similarly the smallest amount of putrefying milk, meat extract or paste added to fresh milk or meat will soon cause the latter to putrefy in the mass.

All these changes differ apparently from those ordinary decompositions which are developed through the effect of chemical affinity. These are chemical actions, conversions or decompositions which become stimulated by substances in the same condition.

It is demonstrable that inorganic compounds made up of the simplest molecules are the more stable and resist decomposition, whilst others more complex in structure are more liable to decompose. The cause seems clear: the greater the number of atoms forming a molecule, the greater the number of ways in which the said atoms are mutually drawn together. Such complex balances are practically unstable and persist only through inertia. In this way, complicated chemical combinations, like the sugars, for instance, become the object of such decompositions or fermentation.

We have seen that some metals acquire the power, not inherent in them, of decomposing certain bodies, such as water or nitric acid among others, through mere contact with other metals in the act of chemical reaction. We have also seen that hydrogen peroxide and other peroxides which decompose easily can transmit this quality to other substances present. This is the same thing that happens, as we may also recall, in the case of fermentation or putrefaction brought about by organic matter which itself is fermenting or putrefying. Different bodies, in the act of combining or decomposing are found to be in a position to induce such processes, that is to cause changes in balance between the component atoms of the molecules, particularly of organic molecules which are highly complex, and causing these atoms to become arranged in other states of balance of greater stability, i.e. in the form of smaller moleculed bodies, in accordance with their natural affinities. Among the substances especially active in this respect must be reckoned in the first place those organic substances containing nitrogen, particularly the albumins.

LOUIS PASTEUR 1822–95

Note sur un Mémoire de M. Liebig relatif aux Fermentations (Note on a Paper by Mr. Liebig relating to Fermentations), Paris: *Comptes Rendus de l' Académie des Sciences*, 1871.

In 1869 Liebig published a long paper on fermentations which has recently been republished in translation in the French *Annals of Chemistry and Physics*. It is an apparently very deep criticism of certain of my researches on fermentations.

In the first of his disclaimers, Liebig formally states I cannot have produced ferment from beer nor furthered alcoholic fermentation in a sweetened mineral medium wherein an extremely small amount of yeast may have been sown. Here is the touchstone of truth or error. Liebig, in fact, holds fermentation, like putrefaction, to be a phenomenon similar in nature to death, if one may say so. Certain substances, and particularly those which he calls albuminoids, such as albumin itself, fibrin, casein and others or the organic liquids containing them, namely blood, milk, urine, etc., possess the property of imparting to the molecules of fermentescible matter the motion which exposure to the air determines in them. This fermentescible matter is divided into fresh products without taking anything away from the inductive substances and conversely without giving them anything either from its own make-up. I maintain on the other hand, that fermentations are correlative with life, and think I have proved beyond cavil that a substance capable of producing fermentation never enters into fermentation unless there is some continuous chemical change going on between its living cells which increase by assimilation of a part of such fermentescible matter.

Liebig's theory was in full swing when I was able to show that in any fermentation special organisms are to be found and that on considering that the phenomena produced were due to the presence of dead albuminoid substances, no account is taken of the fact that life develops alongside and parallel with the fermentation process. I was able to prove, moreover, that all these fermentations become impossible even in contact with the air, if it so happens that the air does not bring right up to the substances in question the organized germs which it ceaselessly carries from one part to another in the strata nearest the earth. Nevertheless, and this is another of my results, fermentescible mixtures which do not ferment, because of the absence of the germs responsible, may be oxidized by mere contact with pure air, suffering appreciable chemical changes which are however quite distinct from those which develop during fermentation.

I made up fermentescible media with the substance to be fermented, suitably selected mineral salts and the figurate germs. Fermentation took place immediately in each case.

And, in accordance with these facts, I assert, for instance, that every time wine becomes turned into vinegar it is through the action of a film of vinegar plant, *Mycoderma aceti*, which develops on the surface. Nowhere on earth does there exist a single drop of vinegar whereby wine may have soured on contact with the air, without the yeast spore mycoderma having been present. This microscopic plant has the property of condensing oxygen from the air, in the same way that platinum black or red blood-corpuscles do, fixing such oxygen in the subjacent materials available.

Liebig denies the correctness of these assertions: "With dilute alcohol used for the speedy manufacture of vinegar, elements nourishing the mycoderma are excluded, and vinegar is produced without its help." Liebig says, moreover, that he discussed this with the manager. of one of the most important acetic acid factories in Germany, Riemerschmied, who affirmed that in his works, alcohol in the course of its conversion received not the slightest external addition and that apart from the air, wooden surfaces of the vats, and coal, nothing is able to act on the alcohol. He adds even that Riemerschmied places no credence in the presence of mycoderma.

Very well then, I would suggest to Mr. Liebig that he gets from the works a few wood shavings, drying them quickly and sending them to Paris for examination by an impartial committee. Mycoderma would certainly be found on the surface of such shavings. I would also suggest a further experiment to him, namely, that he gets Mr. Riemerschmied to lend him one of his vats, which has been in use for some time with such great success, that he puts into it daily an amount of vinegar equivalent to a bit more than half a gallon of absolute alcohol, filling it up with boiling water which should not be left in more than half an hour, after which he should start the vat running again. According to his ideas, it should work just as well as before the scalding, whilst I, on the other hand, maintain that the vat so treated will give no vinegar, not, at any rate, for a fairly long time, that is to say until there will have developed on the inside surface of its staves fresh colonies of fungus, i.e. mycoderma, taking the place of those killed by the boiling water.

SIR FREDERICK GOWLAND HOPKINS 1861–1947

On an Autoxidizable Constituent of the Cell, *Biochemical Journal*, Vol. 15, pp. 286–305, 1921.

Some years ago I was endeavouring to discover if vitamins were to be found among sulphur-containing compounds, and was led part of the way towards the separation of the substance now described. A little later, acting on the suggestion that acid formation in muscle is a necessary factor in contraction I wished to discover if by chance in the absence of carbohydrate, acetoacetic acid from fat might function instead of lactic acid. This led me to apply the nitroprusside test to tissues. At this time Arnold's papers had not yet appeared and I was ignorant of Heffter's publication. . . .

There can scarcely be a doubt that the substance to be described is the "Philothion" of de Rey Pailhade.

In 1888 this author showed that yeast cells and aqueous extracts of yeast have the property of reducing sulphur to hydrogen sulphide. Later he showed that many animal tissues possess the same property. Throughout a long series of communications upon the subject he has courageously maintained the view that the labile hydrogen thus shown to exist in living cells has important respiratory functions. His views as to the probable nature of the hypothetical substance (named as above) which he supposed to carry this labile hydrogen, have been modified from time to time. In his latest writings he speaks of it as the hydride of a protein. After the publication of Heffter, to which reference will immediately be made, he accepted the view that the labile hydrogen exists in sulphydryl groups, HS— . . . .

SUMMARY

A substance responsible for the nitroprusside reaction which is given by nearly all animal tissues, and was applied by Heffter and by Arnold in proof of the presence of sulphydryl groups in the cell, has been isolated from yeast, from muscle, and from mammalian liver. It has the properties of the philothion of de Rey Pailhade.

Evidence is given to show that the substance is a dipeptide containing glutamic acid and cystein. The relation of the two amino-acids in the molecule has not yet been determined. Though present in low concentration (0·01 to 0·02 per cent of the fresh tissue) the dipeptide contains practically the whole of the non-protein organically bound sulphur of the cell.

The substance is autoxidizable, and owing to the changes in the sulphur group of its cystein moiety from the sulphydryl to the disulphide condition and vice versa, it acts readily under varying conditions either as a hydrogen acceptor or an oxygen acceptor (hydrogen "donator"). It can be both reduced and oxidized under the influence of factors shown to be present in the tissues themselves.

Evidence is discussed which suggests that the substance, which we propose to call *glutathione*, has actual functions in the chemical dynamics of the cell.

Reproduced by permission of The Biochemical Journal, Miss Barbara Hopkins and Executors of the late Sir F. G. Hopkins

CHAPTER V

METABOLISM

ASSIMILATION

GROWTH in the animal and vegetable kingdoms is a process from which no individual is exempt. Growth, that is increase in size and weight of the living matter forming the individual, results from assimilation. By assimilation, matter similar to that of the organism itself is built up from the foods consumed, which latter are moulded and converted to form the specific substance of which the whole individual is composed. Despite their general similarity of composition, both the living substance and the greater portion of the organic material in various organisms are different and distinctive, certainly for each particular species and even for each individual within any particular species.

The basic property of live matter is its power of building up substances similar to itself. Consequently, synthesis of living substances always presupposes the influence of some other live matter, and synthesis can only take place when such matter is present. Synthesis comes about in such a manner that fresh living matter makes its appearance complete with chemical, histological and anatomical formations identical to those of the organism entering directly and immediately into the assimilative process. In this identity between the products of assimilatory synthesis and the organized substances which are the instruments of the synthetic process, consideration must be paid to the *specific assimilation* applying to each organism, to each cell thereof, and even to each differentiable portion of such cells. Thus, the growth of an individual with all its chemical and morphological characteristics comes about through progressive assimilation and the number of individuals forming a species is likewise increased.

Assimilation is the process of manufacturing living matter proper, that is to say, the essential cell plasma, residing and living within the cell, whilst at the same time it produces inclusions, paraplastic materials, nutritional reserves or functional products playing a special part in metabolism and in the body functions. In short, through assimilation, nature is able simultaneously to create both the machine and its fuel.

65

These assimilative and anabolic operations are of course endothermic, whence it happens that assimilation must integrate not only matter by itself but energy as well. Typical of assimilative changes are those brought about by the action of chlorophyll. It is known that this substance causes water to combine with carbon to form one of those self-same paraplastic substances just referred to, in this case starch, which forms a food reserve capable of subsequent use for countering the many necessities of the individual. To the substance, in this instance inorganic, namely water and carbon, energy in the form of sunlight has to be added so that organic matter may be synthetized.

Organic molecules, being much bigger and more complex than inorganic particles, contain more heat than the latter. The greater the complexity of the number of atoms forming them, the more the energy within the molecules. Assimilation, the combined whole of the synthetic process, demands fixation of the energy supplied to the material elements making up the large particles which, in the first instance, are organic, and then become alive. It must therefore be borne well in mind that assimilation does not consist merely of the manufacture of complex material edifices or particles of vast structural complexity, but that it must also supply such systems with the amount of energy necessary for their synthesis.

ASSIMILATIVE DISASSIMILATION

The source of this energy is sunlight. Once the glycides are produced, through union of bodies low in energy such as water and carbon, such glycide reserves will be able to supply the energy necessary for further assimilative changes because of their disassimilative decomposition. Energy will pass from a system in disassimilation to another in synthesis, as shown by Pasteur in 1875, when he wrote: "The energy supplied by oxidations typical of disassimilative operations can be used for completing synthesis within the organism." Therefore assimilation and disassimilation develop in indissoluble conjunction one with the other. Coupled catabolic and anabolic reactions frequently occur as instanced by Meyerhof in an interesting example, namely, the resynthesis of glycogen from lactic acid in the muscle, by the use of the energy liberated from a portion of the lactic acid. Such energy of chemical origin being applied to the unconsumed portion of the lactic acid raises it by assimilative operation to a glycide.

Such processes are now known as "the Pasteur-Meyerhof reactions," and occur repeatedly in animal and plant life. Plants, besides being able to convert carbohydrates into fats, also manufacture proteins by incorporation of nitrogen with ternary molecular structures. This stands for synthesis and by the same token a need for energy, which is introduced by decomposition, most certainly oxidation, of a fraction of the glycides resulting from the chlorophyll action. In animals too, the immediate organic principles making up the foodstuffs have to rise to the level of more complex living matter with a greater heat content than the feeding matter itself. This is effected by consuming a part of the food components in disassimilative operations in such a way that they supply the energy necessary for such assimilation. In fact, the three main objects of dynamogenic feeding within the limits of animal metabolism are: maintenance of kinetic energy, generation of heat, and output of energy necessary for assimilatory synthesis.

FUNCTIONAL DISASSIMILATION

Once living matter takes shape, once the peak which is the giant living molecule has been reached, that disassimilative exothermic disintegration commences whereby the vital processes are sustained. The functions get their energetic nutriment from these disintegratory processes which always finish up with a credit balance.

The consumption of dynamogenic matter does not take place solely at the expense of living matter properly so-called, as on the whole it uses up reserves stored within the cells, forming the fuel referred to above. Living organisms, from this point of view, may be compared to a heat engine burning a special type of fuel but also partly consuming certain portions of the engine itself, which portions are immediately replaced. The flow of energy in the case of live organisms cannot be compared to the current of a river as seen from the banks unless we take it that a portion of the banks themselves are carried away *by* and *in* the stream itself at the same time.

LIVING MOLECULES IN THE STATIONARY STATE

Metabolism, an essential feature of life, is a continual self-destruction and self-recomposition of living matter. Cuvier called it a "living whirlpool." Goethe before him had written "life is a flame," whilst, according to Buffon, it was "a minotaur devouring its own substance." It is in fact a constant flow of matter and energy; living systems are complexes in a stationary condition (Ostwald), seat of continuous changes of matter and energy under the cloak of stability.

Pflüger in 1898 portrayed the living molecule as a figure in three-dimensional space made up of rays darting upwards, downwards, rightwise and leftwise, forward and backwards from a common centre reaching towards the surface of an ideal sphere, such rays being composed of groups of radical atoms corresponding to the various immediate principles, proteins, glycides and lipids forming an extremely complicated particle which has a very high molecular weight despite the low atomic weights of its separate elements. To this concept of the living molecule Verworn gave the name *biogen*.

From the biogen linear atomic groups of greater or less complexity, proteins, carbohydrates or fats, or portions of such radicals, emerge into the open, these being the "lateral chains" which Ehrlich in 1904 sought to use as an explanation of the phenomena of immunity. These lines of atoms ceaselessly falling from the biogen into the disassimilative processes would leave atomic remnants in the remainder continuum of the biogen, upon which by assimilation further chains would build up similar to those already broken away. We must imagine the biogen, therefore, as the seat of constant renewal of its constituents, some disappearing whilst others make their appearance, and that, despite all this, the chemical identity of the whole is preserved. There takes place in it—and this is the essence of life—an uninterrupted flow of matter and of energy; assimilation makes up for losses caused by disassimilation. Disassimilation liberates energy whilst assimilation on the contrary demands it. Thus it is that the stream of matter in metabolism entails a corresponding flow of energy.

Bertalanffy (1933–1950) names the aforesaid uninterruptedly continuous physico-chemical vital process an "open system." The organism is a system or combination of unified systems in an open steady state of rhythmic constancy, to which the living being tends ever inexorably to return whenever disturbances become apparent. This rhythmic recurrent balance is a basic characteristic of existence, and when it is interrupted, life ceases.

Biogens of various types go into the formation of cell plasma, cells, tissues and organs. Biogens are arranged according to their physico-chemical structure, and consequently result in cytological and histological structures which make up the complete anatomy. Nutrition is a property of biogens and therefore of the cells. Living matter attains its animate condition through its nutritive capacity; metabolism must be therefore one of the principal factors in life.

Gautier wrote in 1892: "In the structure and organization of chemical molecules forming the protoplasm and in their manner of association one with the other lies the fountain-head of the whole range of elementary phenomena of life. Variation of the integrating molecule causes changes also in the style of reaction throughout the whole organism; vital processes are the direct consequence of the chemical functions of the constituent particles."

METABOLIC MECHANISMS

Assimilation and disassimilation are so bound up with one another as to form a single feature. The former was a subject of research long before the latter. Knowledge of the existence of certain processes of assimilation goes right back to ancient times, whilst disassimilation is on the other hand a comparatively recent discovery. Nevertheless greater attention has been paid to the study of disassimilative processes than to acts of assimilation. Both these processes are equally important in forming the metabolic whole. They are of equal effect in the state of nutritional balance; anabolism overrules catabolism during growth, and conversely, disassimilation preponderates in all the aspects of organic impoverishment.

Disassimilative changes present nothing which is exclusive and particular to living bodies. As we know, Lavoisier taught that breathing is exactly the same thing as combustion. Hydrogen and carbon are consumed inside living organisms in exactly the same way as in any other combustion process. Living matter disintegrates and decomposes into smaller particles during disassimilation as it does in death. Disassimilation is like bodily decomposition. In both cases the breaking down operations are exothermic and in disassimilation there resides a store of energy available for the functions. Thus Claude Bernard was able to say: life, that is function, is death. Catabolism, like putrefaction, is an act of decomposition, simplification, a tendency to change back living or organic matter into inorganic matter. Such changes presuppose a positive thermal set-up both inside and outside the organism.

Assimilation, on the contrary, occurs solely in living organisms and is specific to them as we have already seen. Le Dantec in 1904 catalogued the characteristics of the disassimilative processes as compared with the assimilative, putting them into the form of two basic formulae of nutrition. In disassimilation, which in general circumstances hardly differs from any other disintegrative or analytical change, the substances reacting in the first term of the equation are not to be found in the second, a total conversion of substance having taken place both in the nutritive material and in the cell components.

$$N + O = R$$

i.e. Nutriment plus Oxygen equals Residues.

In the assimilation formula, one of the reacting terms, the living matter, is found generally in increased amount in the second term—

$$N + L = nL + R$$

i.e. Nutriment plus Living matter equals n times the Living matter plus Residues.

The object of assimilation is to build up living matter by consumption of foodstuffs. That is why, therefore, foods are divided into respiratory—as they were called in former times—or dynamogenic, and plastic or morphogenic; in other words, foods for heat-generating combustion and foods for the formation of living matter. These two facets of metabolism would not occur independently. Claude Bernard himself pointed out that prior disassimilation is a factor of assimilation. To repair a breaking off of portions from the biogenic particle, obviously such amputation must first have taken place. As we have shown, it might be thought there was some contradiction between assimilation and function, but elementary observation proves the contrary, namely, that an organ in operation continues to develop, as for instance a muscle. Rodrigo Lavin in 1900 emphasized this when he asserted that assimilatory synthesis is a functional phenomenon. In a function, of which the muscle is the most characteristic example, food reserves are consumed mainly as a supply of fuel for the production of energy. Living matter as such takes part in the chemical changes demanded by the function and that entails an increased biogenic disintegration. The replacement of molecular complexes destroyed by the functions is more than taken care of by reparatory mechanisms, from which we get an increase in the synthetic operations, and an augmentation of the living matter accompanying and distinguishing the function.

ANOXYBIOSIS AND OXIDATION

Disassimilative disintegration was considered a few years ago to be exclusively of an oxidizing character, according to the ideas expressed by Lavoisier. Our increased knowledge of facts now shows that the chief part in disassimilation is taken by anaerobic operations without the aid of oxygen, these being the chemical reactions labelled by

Pasteur as fermentative. Oxygen then acts upon the products of anoxybioses whilst combustion completes the acts of disassimilation. This twofold aspect of the process was brought to light mainly through Gautier's researches in 1898 and has more recently met with repeated confirmation.

Disassimilation takes place through the agency of various diastases leading the living matter or reserves through all stages to excretion. Assimilation is similarly the work of synthetizing diastases, i.e. enzymes, whose exact operation is at the present time even less known than that of the catabolizing diastases. Each of these two types of diastase regulates its activities in such a way as to ensure exact adjustment of the intensity and quality of nutritional changes to the vital needs of each individual, be it plant or animal.

The result of this is that the intensity of nutritional changes comes about neither through chance nor even by reason of special substances at the disposal of the living matter in its uninterrupted course of metabolism. Living matter automatically regulates the intensity of chemical changes taking place in it, by means of properties which we must perforce call vital, intrinsic in itself, from the very fact of its being living matter.

This impelled Pflüger to say in 1905: "The live body cannot be compared to a fire which burns all the more fiercely the more fuel is thrown on to it; quite the contrary: whatever may be the quantity of material, it oxidizes only the exact amount needed for its functional needs. If the materials of combustion are excessive in amount, they stay stored up in the form of fat. Consequently, what determines the total of work in our organs is not the amount of matter supplied by the food, but the quantity of material actually used."

To which Bouchard added in 1906: "A portion of body albumin, eight parts in a thousand in young people and four in a thousand in older folk, is probably destroyed every day and replaced by an equal amount of albumin derived from the food consumed. The amount of living albumin renewed comes out practically constant for each stage in life and for each set of circumstances. It varies according to age, decreasing as time rolls on; it increases during certain illnesses particularly when fever is present and decreases in others, especially in certain diatheses. It is not affected by muscular effort nor by cold, since the foodstuffs supply—quite apart from the albumin which has to replace that lost in the tissues—sufficient organic matter to liberate the calories required by the effort or by the struggle against cold. Neither is it affected by variations in the intake of oxygen or protein foods, since such foods are ingested in sufficient quantity."

Nutritional changes develop therefore in all forms of animal and plant life as the basis of some form of vital manifestation, animal or vegetable. Disassimilation is particularly manifest in animals, which move and, on getting heated up, require calories which are in fact supplied by the disassimilative, exothermic operations. Apart from such differences in quantity, the metabolic mechanisms in animals and plants—not counting the chlorophyll action—are identical.

*   *   *   *   *

CLAUDE BERNARD 1813–78

Leçons sur les Phénomènes de la Vie, communs aux Animaux et aux Végétaux
  (Lectures on Reactions common to Animals and Plants), Paris:
  Baillière, 1878.

I consider that two types of *phenomenon* develop necessarily in the living body—
  1. The *phenomena* of *creation of life*, or organizatory synthesis.
  2. The *phenomena* of death or *organic destruction*.
It seems to me desirable to explain the meaning of the terms *creation* and *organic destruction*.

If in regard to inorganic matter it be admitted and rightly so that nothing is lost and nothing is created, nevertheless in the case of living beings it is quite another matter. Everything is created morphologically, becomes organized, and then dies or is destroyed.

In the developing embryo, muscles, nerves, bones, etc., appear in the spot corresponding to their place in the organism repeating the form of the individual from which the egg was generated.

The surrounding matter is assimilated to the tissues either as a nutritive principle or as an essential component. Thus is created the organ which can be distinguished by its structure, shape and properties.

On the other hand, the organs are destroyed and disorganized continuously and precisely through their own functioning. Such disorganization constitutes the second phase of the great drama of life.

The first of these two orders of phenomenon is individual and special to the living body. This evolutionary synthesis is the very essence of life. I will repeat on this occasion the expression to which I gave voice some time ago, namely: "Life is creation."

The second order of phenomenon, the destruction of life, is of a physico-chemical nature, normally the result of combustion, fermentation, putrefaction or some other action, comparable in short to a large number of chemical acts of decomposition or disintegration. These are true phenomena of death when they take place in an organic being.

And it is worthy of mention that we delude ourselves habitually in such form that when we want to point out phenomena of *life*, what we really talk about are phenomena of *death*.

Vital phenomena properly so-called do not attract our attention. Organizatory synthesis works silently within the body, its actual occurrence well hidden away from sight, gathering up secretly the materials which will afterwards be consumed. We take no account directly of these phenomena of organization and assimilation. Only the histologist or embryologist studying the development of tissues, cell elements or the living body as a whole becomes aware of the changes and phases which reveal this hidden working; in one case it may be the deposit of material, somewhere else it will be the formation of a skin or a nucleus, further on still a multiplication of cells or a renewal of some organ or other.

The phenomena of destruction or death and disassimilation are the most obvious to our senses and seem therefore to us to characterize life. Their signs are clear; a movement takes place because a muscle must have contracted as a result of some corresponding wish; when stimulated a gland secretes. It is the substance of the muscle, nerve, brain or gland which is disorganized, destroyed and consumed. And this is how every living manifestation combines with an organic destruction and that is what I meant when I wrote the paradoxical phrase "life is death."

The existence of all animals and plants is maintained by these two types of necessary and inseparable processes: *organization* and *disorganization*. Our science has to fix the terms and conditions which guarantee the correspondence of these two types of phenomenon.

This division of vital phenomena which we have adopted—*assimilative* and *disassimilative*—is an exact expression of the truth and comes about simply from proper observation of life phenomena. But life is the combination of both types of manifestation.

CHARLES BOUCHARD 1837–1915

Cours de Pathologie Générale (General Pathology), Paris: Masson, 1879.

A common property of all living particles is a particular molecular movement which is not observable in either inorganic or dead matter. The effect of this movement is to introduce into the living particle certain extraneous non-living substances and submit them to chemical metamorphoses which I will call *vivifying*, since by means of such changes the said substances come to form a constituent part of the live element and participate in its life; there are also other changes which I would call *retrograde*, whereby living matter ceases to exist and is converted into products of decomposition which have to be expelled from the organism. The distinctive feature of the activity of any living particle is, therefore, a twofold continuous molecular movement of introduction and expulsion simultaneous with a double operation—also continuous—of chemical transmutation, the one following the introduction of substance and the other preceding its ejection. On the one hand we have translation, and on the other transmutation: a double physical operation simultaneous with a double chemical effect.

If nutritional material conversions are a common characteristic of anything alive, if nutrition is therefore the general characteristic of life, the best definition of life itself will be that given by Aristotle: "Life is the combination of the operations of nutrition, growth and destruction." This idea corresponds with what Blainville set out in other terms: "Life is the double internal motion of composition and decomposition which at the same time is general and continuous." Nutrition therefore —we may repeat—is life; it is life with its double motion of assimilation and disassimilation, of creation and destruction at the same time.

Assimilation and disassimilation are in general parallel and simultaneous but not invariably so, not even in the physiological state. Growth is the predominance of assimilation over disassimilation. In pathological conditions this predominance is the cause of over-development or hypertrophy, whilst atrophy and cachexia come about through excessive disassimilation. And we may affirm withal, that even when the physical and chemical actions—each of them twofold in the sense of being integrative and disintegrative—are not found to be indissolubly bound together between themselves at every particular moment, they are jointly and severally equally necessary for the maintenance of life. None of them may be hampered or suppressed without the nutritional molecular motion being held up and life being consequently extinguished.

Reproduced by permission of the publisher

ARMAND GAUTIER 1837–1920

La Chimie et la Cellule Vivante (Chemistry and the Living Cell), Paris: Masson, 1894.

Life makes its appearance in the cell and passes from generation to generation thanks to specific substances transmitted from parent to offspring and determining the organic forms and molecular structure whence the elementary functions are derived.

The regular functioning which preserves the individual and perpetuates the species consists of a series of orderly acts in whose fulfilment every cell plays its special part. Each lives, however, more or less independently according to its structure, assimilating nutrition and embodying it in the mould of its own specific constitution.

We do not know the method of molecular differentiation and organization for each species of cell. All we know is that the parts which govern and specialize the working of the cells are the nucleus and the protoplasm, each made up of specific albuminoid substances. The nucleus rules the activity of the protoplasm and its chief importance is the maintenance of morphological type and the development of the cell. Protoplasm works and assimilates its surrounding medium thanks above all to its plastids or specific granulations, which are differentiated organs proper to each species and frequently varying in nature even within the same cell.

Animal cells cannot from their elements synthetize the albuminoid ingredients of their protoplasm. The proteins arriving within them, or perhaps their more immediate derivatives, are converted into particular albuminoids due to disintegrations, syntheses, and isomerizations which modify, rejoin and assimilate simpler materials absorbed in the digestive tube or arriving from other cells.

In the living cell-protoplasm fundamental protein substances are disassimilated mainly by hydrolysis in a reducing medium and sheltered from any influence of oxygen. From this arises the formation of fresh nitrogenous substances, amides or complex amines, which are definitely converted into urea, carbohydrates and fatty bodies by the effect of simplifying anaerobic reactions.

Urea, the definite form in which some fourteen-fifteenths of the total nitrogen in the tissues is eliminated, seems to be produced in the main independently from the oxidation processes. Between this substance and the protoplasmic albuminoids we have a whole range of intermediate nitrogen compounds. The most complex are proteins still full of considerable activity, such as peptones, diastases and toxins. Afterwards we get creatinic substances and ureides which immediately precede the formation of urea itself. Furthermore, there are xanthine compounds and uric acid arising from nuclear materials. Many of these derivates are eliminated through the urine or bile whilst others serve for transient syntheses or serve to stimulate digestion or even nervous activity.

The ternary products, formed by disassimilation of the protein substances in correlation with urea and other nitrogenous bodies, vanish through oxidation controlled by the oxidases. The sugars are consumed or changed into fats by loss of carbon dioxide. The fatty bodies are saponified and then oxidized by degrees. It is precisely these oxidations of sugars, fats and other ternary compounds which are the main source of energy for the warming-up and work of the organism.

The chemical phenomena, developed in the cell, supply the energy necessary for every manifestation of life. The organization determines merely the direction which this energy must take and the order of the phenomena which preserve the cell, the individual and the species.

Reproduced by permission of the publisher

OTTO MEYERHOF 1884-1951

La Relació entre els Processos Fisics i Quimics i la Contracció Muscular (The Relation between the Physical and Chemical Processes and Muscular Contraction), Barcelona: Institut d'Estudis Catalans, 1934.

The application of purely chemical methods to the study of the mechanism of muscular contraction has given us the following results:

1. Oxidization of the foodstuffs is not the process directly supplying work, but serves solely for the purpose of replacement. Contraction

may take place even in the absence of oxygen, at the expense of the energy liberated by anaerobe break-down processes, in which case the decomposition of glycogen makes way for the formation of lactic acid.

2. Nor is lactic acid formation the process directly liberating the energy required for contraction; since, as Embden demonstrated, such formation takes place, to a large degree, when contraction has already ceased, and Lundsgaard's experiments showed that through intoxication of the muscle by monoiodoacetic acid, the formation of lactic acid comes to a full stop, and the muscle is then still able to carry out a certain amount of anaerobic work. Under such conditions, creatine phosphoric acid, discovered by Eggleton and Fiske, decomposes more intensely than usual and it is this decomposition which supplies the energy necessary for contraction. Whilst in a normal muscle, the formation of lactic acid starting from glycogen, serves partly to resynthetize the creatine phosphoric acid broken down, just as in respiration, oxidization is employed for resynthetizing the lactic acid to glycogen.

3. It is quite likely that neither does the decomposition of the creatine phosphoric acid directly supply the energy required for contraction. At least that is what Lohmann's latest experiments lead us to think. Here again, it may be a matter of a process of recovery whilst some other chemical conversion—decomposition of adenylpyrophosphoric acid into adenylic acid and phosphoric acid—is taking place at the same time as the contraction and appears to be reversed during relaxation.

The idea that the breakdown of adenylpyrophosphoric acid might take place prior to that of the phosphocreatine is due to Lohmann's observations verified not only in intact muscles but also in muscular extracts. We have already been aware for some years that *coupled chemical reactions* take place in such extracts and must be borne in mind when changes in energy in the muscle are under consideration.

The following facts can be established in connection with healthy muscle and its histological structure. The break-down of creatine phosphoric acid is impossible unless adenylic acid and pyrophosphoric acid is simultaneously synthetized to adenylpyrophosphoric acid. It is therefore necessary that the first reaction to take place after excitation should be the break-down of adenylpyrophosphoric acid into adenylic and phosphoric acids. Only then will it be possible for the phosphocreatine to split up so that the adenylpyrophosphoric acid previously decomposed now becomes resynthetized. If the decomposition of the adenylpyrophosphoric acid takes place in between the instant of excitation and the breakdown of the creatine phosphoric acid, we may assume, in agreement with Lohmann, that such reaction takes place simultaneously with the mechanical process of contraction.

Reproduced by permission of the publisher

ARCHIBALD VIVIAN HILL *b.* 1886

Alguns Avenços recents en la Fisiología del Muscul (Some Recent Advances in Muscle Physiology). A lecture delivered during the Course in Biochemistry and Physiology of Muscular Contraction and Fermentation at the University of Barcelona (Curs de Bioquímica i de Fisiología de la Fermentació i la Contracció Muscular—Institut d'Estudis Catalans), 1934.

Muscular contraction can be analysed into the four following stages:

First stage: in which the response is produced

Second stage: lasting throughout the whole period during which the stimulus remains active

Third stage: when the stimulus ceases to act and the muscle relaxes

Final stage: when, under the influence of oxygen, the muscle recovers from the changes taking place during the period of activity

An emission of heat is observable during the rising mechanical response, and a continued production of heat, becoming constant in rate during the period in which contraction is maintained. Coincident with relaxation, the production of heat first diminishes, afterwards again increasing, and finally decreasing to zero as relaxation terminates. The *heat of relaxation* represents the disappearance of the potential mechanical energy necessarily present in the muscle in isometric contraction. One day we shall learn what the chemical reactions are which are associated with each one of these stages.

The anaerobic heat curves can be considered as the energy outline of chemical reactions taking place in the first half minute after contraction, as discovered by Embden, Meyerhof and other investigators.

An interesting fact, not yet explained, is that the maximum intensity of the delayed heat comes first and is more intense after a long stimulation than after a short one. Such a difference in the form of the *delayed anaerobic heat* appears with such regularity and so plainly that there must be some reason for its occurrence. The amount of this heat is not the same in subsequent contractions: as the muscle gets more and more tired the delayed heat diminishes relatively more than the initial heat which it follows.

This delayed heat is liberated also in the presence of oxygen, but to it is added the heat of recovery which has its source in oxidation and is of longer duration. The ratio between the total heat (produced during and after contraction by a muscle to which sufficient oxygen is supplied) and the heat first liberated (that is, during contraction) gives us important information about the chemical mechanism of contraction. The total heat set free in the quarter-of-an-hour or twenty minutes during and after contraction maintains a ratio to the initial heat of contraction which was originally thought to be absolutely constant. This is not exactly so. In the case of a relatively long tetanus the ratio is about 2·4; in a short tetanus, or in a series of twitches, the figure falls to about 2·0; and in an isolated twitch, the ratio lies between 1·6 and 1·7. From such facts, it may perhaps be deduced that recovery is not invariable, but depends on the nature of preceding activity. The *efficiency* of recovery is greater in the case of a simple isolated twitch than in that of a tetanic contraction. Perhaps, some day, Professor Meyerhof, whom I see here listening to me, and who has contributed so much to the advance of our knowledge of the chemistry of muscular contraction, will be able to tell us why things happen this way!

A curious phenomenon, discovered by my co-worker, Dr. Feng, is that the metabolism of the muscle at rest increases as the load increases. Taking the load away causes the resting metabolism to drop back immediately. The same effect can be seen on measuring the consumption of oxygen by the muscle.

Reproduced by permission of the author

CHAPTER VI

GROWTH AND REPRODUCTION

CELL DIVISION AND GROWTH

ALL cells have a certain limit for size, not varying appreciably from one to the other and measurable in thousandths or ten-thousandths of an inch. Of course some cells are larger than others but the differences in size are not really very great. On the other hand, however, we do find giant vegetables and animals living in the same environment as microscopic and practically invisible creatures and plant organisms, all of them from the least to the greatest containing a huge number of cells. The fact remains that the size of the cell element is practically constant, and that of the individual depends on the number of cells forming it, this being a characteristic index for any particular species.

Again, in reproduction through germ cells, the whole of the organism derives from progressive division of the originating cell, that is a zygote in the case of sexual reproduction or some other cell in place of the zygote when reproduction is asexual. From the division of such cell we get two blastomeres which split into four and these again into eight, then into sixteen, thirty-two, and so on to infinity. In this way the embryo keeps on increasing through multiplication of its cells, whilst after the embryonic stage growth still continues until the individual reaches adult size, its number of cells increasing all the time. Growth is therefore in all its aspects a consequence of cellular multiplication developing progressively until the volumetric limit for the species in question has been reached.

Thus we see why growth cannot go on indefinitely; it must cease as soon as the specific limiting size has been reached, and it is at this point generally that the reproductive faculties of the individual make their appearance. Reproduction becomes thus a continuation of the growth process. There is, indeed, an increase in mass in species because the number of individuals increases, but no single one of the species increases beyond its particular limiting volume, and when that is reached individual growth ceases. The division of its cells *then* serves only to replace those cells which have died off or, in the case of natural or artificial mutilation, to repair such damages.

Cell division is a distinguishing feature of living matter, a prime factor like nutrition and indeed a consequence of the latter. When assimilation is playing the major part, the cell grows, but there must be some fixed relationship between volume and surface area, since it is through the surface that interchanges of substance and energy with its environment take place. Increase in size of cell implies mastery of volume over surface and decrease in their ratio, which means that a moment will come when the surface will be insufficient for the continued existence of the cell; this is circumvented by division of the cell, which when it has become sufficiently large breaks down into two or more.

Schleiden and Schwann (1839) assumed that the cell was the offspring of the plasmic fluids, the blasteme, brought into existence in much the same way as a crystal appears in a saturated solution, by precipitation of the nucleolus followed by the nucleus which latter is surrounded with protoplasm. In 1854, Nägeli for the first time established that one plant cell is always born from division of another, whilst later on, in 1858, Virchow confirmed that this takes place in all classes of cell whether plant or animal and epitomized this in his well-known saying "every cell comes from some other cell."

TYPES OF CELL DIVISION: AMITOSIS AND MITOSIS

Division of the cells can be effected by different processes. The simplest is that of direct splitting to be observed in single-celled organisms such as protophytes and protozoa, and also in some cases metaphytic and metazoic tissues. Ranvier in 1886 verified the direct division in leucocytes and still later the same thing was confirmed in cells of other animal tissues, in the connective for instance.

In direct division or amitosis, the mass of the cell keeps on lengthening at the same time becoming constricted until the nucleus divides into two parts. This division of the nucleus is followed by excision of the protoplasm into two portions each surrounding one of the two halves of the nucleus. This is how the two cells resulting from partition of the mother-cell are formed and the process may be repeated through a long series of generations.

More frequent than this direct or simple division is another form of cellular multiplication, in which changes characteristic of the morphology of the nucleus take place. This is karyokinesis, so called by Schleicher in 1878, or mitosis found by Flemming, also in 1878. Its details were given by Strasburger between 1876 and 1884, and

Flemming showed that it also takes place in animal cells. The nuclear chromatin is arranged in threadlike form, the spireme in the prophase, and this breaks down into a fixed number of sections, the chromosomes, which after curving into hairpin shape form a sort of star in the equatorial plane of the cell, around the achromatic spindle which joins the two protoplasmic centrosomes. Each of these shapes splits longitudinally into two, and thus the combination thereof, the mother-star or chromatic-aster, divides into two daughter-stars with the same number of chromosomes in the anaphase. The daughter-stars are attracted as it were each to its respective centrosome and then move away from the equator of the cell along the achromatic spindle, this being the metaphase. Having once reached the centrosomes, the daughter-stars become the seat of an inverse process to that which fostered the individualization of the chromosomes; this completes the telephase. The forks of the daughter-stars lose their inflexion and degrade into straight lines, the various resultant sections becoming joined together in spiremes and very soon in these the morphological signs of reproductive activity disappear and they now become the chromatic formations of the daughter-nuclei. Around these chromatic formations the respective nucleus is built up again, and around each nucleus the protoplasm is divided, so that, as in the case of amitosis, one cell has become two, its fund of nuclear chromatin being shared symmetrically and equally between the resultant daughter-nuclei.

Mitosis does not always occur with the regularity just described. It must be pointed out, in the first place, that there are certain multi-polar mitoses where, instead of two centrosomes and one achromatic spindle, three, four or more centrosomes make their appearance with their corresponding spindles almost always intertwined and their mother-stars in a plane perpendicular to that of the axis of each one of these spindles. Understandably in such cases the cell will split into three, four or more according to the section planes dependent on the spindles and this practically always in an irregular manner. However, in some of these mitoses the cellular division will not be completed and the resultant cell may contain several nuclei. These non-typical or substandard mitoses are particularly to be observed under patho-logical conditions, for instance in certain cases of poisoning (Hertwig, 1887), in tissues of abnormal growth, cases of irritation, neoplasms (Galeotti, 1893) and others, also in the case of polyspermic fertilization.

Though mitosis is the most typical way in which cell division takes place in the tissues of metaphytes and metazoa, this does not mean that direct or amitotic division may not occur in such organisms, and in fact this may be so. Direct division however most frequently occurs in the more simple organic forms such as protophytes and protozoa. It is possible, moreover, to observe cases which might be held to be intermediate between amitosis and mitosis, for instance nuclear chromatin counterparts which, though not characteristics of karyokinesis, are more complex than those developed in simple division and recall certain nuclear aspects of mitotic division. There is room for intermediate gradation between the two forms of division of the nucleus and of the cell (Della Valle, 1911).

THE SIGNIFICANCE OF MITOSIS AND AMITOSIS

The significance of the two ways of producing multiplication of the cells has been the subject of much discussion. It is generally assumed that mitosis represents some form of improvement, or a more advanced way of doing things. It takes place in the fertilized female gamete, the zygote, and from this is derived an argument in favour of the superiority of karyokinesis. This opinion does not, however, enjoy the widest acceptance. Certainly amitosis is found mostly in reproduction of the more simple forms of life, but there do exist cases of amitosis in very active young cells of metazoa and highly differentiated creatures, which allow expression of the opinion—contrary to what is generally assumed by the majority of writers—that direct division may be a manifestation of an advanced degree of specialization, and that by no means can it be held to be a process of the most elementary simplicity.

It seems, however, that within a general framework, amitotic reproduction may be the more easily seen in the rudimentary forms of cellular life than in the more differentiated. Still, retaining such framework, the appearance of one or the other reproductive variety may depend on physiological conditions and special circumstances ruling at particular moments of existence.

Within the general biological framework above indicated we should have to debar the idea of any determinism, from causes still unknown, deciding the form of the process of division. In short, though it be safe to say that amitosis prevails in the lower organisms and karyokinesis in others of higher botanical or zoological lineage, this does not mean that the ruling plan is rigid and inevitable, but that different circumstances might modify it and promote consequently either amitosis or mitosis whether in animal or plant according to immediate circumstances.

We must point out also the possibility of multiple cellular division. Various nuclei are formed by plurality of division of the original nucleus, and each one of these nuclei becomes surrounded by the corresponding part of the protoplasm, the corresponding cells thus being born (conitomy). This is a process of reproduction observable in certain protozoa. A similar mechanism is that of sporulation. In other cases a species of nuclear gemmation occurs; spots of chromatic material accumulate in the surface of the mother-cell, arising from expansion of the nucleus in such cell, and each one of these spots becomes the centre of a new cell.

There are obviously various ways in which cells may divide, but the more usual processes are in any event direct division and karyokinesis.

EMBRYONIC DEVELOPMENT AND REGENERATION

Such mechanisms of cellular division assure the development of embryo and subsequent growth. Such development consists in multiplication of the number of cells and simultaneous differentiation thereof. Starting with a single cell, a huge number of them are formed in various tissues as parts of the different organs. Multiplication and differentiation of cells are particularly active during the embryonic stage, but continue also throughout life, though of course with ever-decreasing intensity from stage to stage. Differentiations are however still possible even after the embryo has developed.

In certain cases the formative process regains its activity and in mutilations, for instance, there is the possibility of certain organs and even the complete individual being reproduced again. The ability to make good varies from species to species ranging from rebuilding of complete organs and even of a major portion of the whole organism down to the simple cicatrization of wounds.

Division of the cells and differentiation arise from a first assimilative cellular impulse for growth and multiplication, though at the same time the presence of other cells and the influence of the whole being or organism play their part as may be seen in the case of tissue cultures. The cultured cells tend generally to become dedifferentiated (cataplasis), which Champy in 1915 attributed to lack of action on the part of the organism as a whole. The vegetative power of the protoplasm belonging to the cells prevails over the specific differentiating power of the whole with its various parts. Thus it is that Gracianu was able to say in 1931 that cellular proliferation in the test-tube, in cultures, differs entirely from what takes place during the normal evolution of an embryo.

Such influence is exerted by means of various substances, such as trephones, auxins, growth hormones, biotin and others, present in extracts of various tissues and more particularly in young and embryonic tissues. These substances revive the action in cultured cells and are probably the same as those promoting normal development.

The agents of such development are provided with a high degree of specificity and profound differentiations govern the formation of particular organs, or "organizers" as Spemann and Mangold, for instance, called them. Combined action of all the growth factors moulds the ontogeny of an organism and the possibility of repairing mutilations is included in this, more or less in proportion to the phyletic degree at which the creature, or even a plant, may stand. It is natural, therefore, that the capacity for such repair of damage will not be identical at all parts of an organism.

REPRODUCTION BY REGENERATION

In the vegetable kingdom this capacity for reproduction through regeneration of mutilations is usually unlimited, as for instance in grafts. It is common knowledge that many plants multiply by thrusting into the ground fragments of shoots or branches or in some species even leaves. From the underground portion radicle organs are born and from the top part the breathing system consisting of branches, buds, leaves, and flowers. Plant species can also be propagated by means of roots and likewise by sticking the ends of bent-down branches, and even cuttings and so forth, into the ground. Grafts are of a similar nature to these latter. A shoot from one plant engrafts itself into another in such a way that the cambiums of each of them make contact, thus establishing mutual chemical effects between the cells of the graft and those of the host, whence modifications may occur in the host varying in nature and extent according to the particular case. Such modification impressed on the form and functions of the host can become extremely pronounced.

Regeneration with the formation of new, separate individuals is observable also in particular species of animals. In 1774, Trembley showed that after repeatedly taking sections from a hydra, a new hydra could be formed again from each one of the parts cut off. Up to the eighteenth century it was held that such regeneration was a special feature of the plant world and hence it was argued the hydra must also be a plant. It was not, however, very long before it became possible to show that other creatures, of much greater differentiation than the hydra, were also blessed with the same faculty.

Certain worms, particularly the flatworm Planaria *Linnaeus socialis*, the earthworm and others, represent remarkable instances of this. Some echinoderms and starfish are able to build themselves up again, starting with the minutest fragments down to a thirtieth part of their body. As the organs become more complicated the regenerative faculty becomes correspondingly less, but some well-known examples are worthy of note, such as the possibility of forming anew the heads of snails, the nippers and eyes of crabs, and the tails of salamanders, tritons and newts, which latter are already among the vertebrata. Many other examples are forthcoming among various articulated creatures, molluscs, tunicates and the rest. In these processes of regeneration, the first phase is that of dedifferentiation of the cells, i.e. almost as if the cells returned to their embryonic state, this being followed immediately by morphogenous differentiation.

Regeneration is in evidence right from the simplest single-celled creatures and among many protozoa, as shown by Gruber in 1885, Balbiani in 1892 and Nussbaum in 1908 with their researches into merotomy. For the cell to become reconstituted it is essential that the broken-off fragment shall contain a portion of the nucleus, which latter governs the process of regeneration. In polyplastids such regeneration does not depend on the reconstruction of a fractionated cell but on cellular division as we have just seen. In the division of the cells the main part is again played by the substance of the nucleus.

REPRODUCTION THROUGH SOMATIC OR BODY CELLS

Another process of reproduction by somatic cells is that known as gemmation, which may well be considered a type of genetic or organic growth. Gemmation appears in single-celled beings, and is frequent for instance in the saccharomycete cells and among plants. Undifferentiated groups of cells are able to generate the entire plant, as happens with the buds of the foliate appendages of ferns, and the inflorescent bulbils of *Allium*, *Lilium*, *Ficus*, etc., and the buds or sprouts of numerous aquatic plants and the like.

In the animal kingdom the most prominent case is that of the hydras and hydroids, polyps and jellyfish. In the body surface of the individual there appears a bud or excrescence which gradually assumes the shape of the individual and finally becomes detached, constituting a new hydra by simple growth. Certain worms also can reproduce through gemmation, sometimes through successive regenerations. In certain squirts, ascidia and other tunicates living fixed on to rocks, expansions occur in the basal regions or stolon, at the ends of which buds develop, each forming a fresh individual.

Strobilation or transverse fission is similarly a type of gemmation. The creature places various transverse portions one on top of the other which remain joined together for a time and then become successively separated layer by layer, starting with the first, each layer giving rise to a fresh individual. This happens particularly among the coelenterata of the scyphomedusa group, which divide by this means into different individuals.

In all these cases, reproduction of plants and animals comes about simply by multiplication of the cells forming the tissues, the somatic cells, and no special factors such as germ-cells enter into the question. Reproduction is in these cases nothing more than regeneration or, as we may call it, "growth."

* * * * *

EDUARD ADOLF STRASBURGER 1844–1912

Das Botanische Praktikum (Practical Botany), Jena: Fischer, 1884.

The most suitable living object for the purpose of studying mitotic or indirect division is the cell, together with its corresponding nucleus, of the spiderwort *Tradescantia virginica*, though observations should be made at a stage of active development prior to the completed formation whilst the cells are still multiplying in large numbers. For these investigations flower-buds measuring from a fifth to a quarter of an inch in length without their stalk should be selected, and their anthers removed by means of a fine forceps after opening up. Next, the ovary and filaments under the insertion should be cut transversely with a scalpel and separated from the other flower organs, and then placed in a 3 per cent sugar solution on a slide so that the filaments may be examined under an ordinary microscope.

The resting nucleus of the cells forming the hairs looks as if it were

finely stippled, though under a stronger magnification it will be seen
that the minute granules forming it are not isolated one from the other
but connected closely together in rows of fine coiled threads, which
permeate the whole nucleus and form a network or framework entirely
surrounded by a delicate membrane. Nucleoli of various sizes can be
made out between the coils. The nucleus is surrounded by the peri-
nuclear cytoplasm connected with the peripheral cytoplasm beneath
the cell membrane by means of a number of irregular cytoplasmic links.
In addition to the hardly distinguishable microsomes, the cytoplasm
includes larger, more strongly refractive grains, some with products of
cellular existence, the leucoplasts.

When getting ready for division, the nucleus swells up whilst its
chromatic framework shapes into a coarsely granular thread which
then commences to lengthen, whilst the coils tend to become arranged
roughly parallel to each other. These progressive changes are easily
seen in one and the same cell, but such observation requires a comparatively long time and must be maintained continuously or interrupted only for brief intervals. The grains gradually assume a homogeneous appearance and at the same time the nuclear membrane
disappears. After a while the swollen thread breaks down into
segments or chromosomes, which then form into straight rods
approximately equal in length, arranged in two groups one above the
other with their equatorial ends meeting and their polar ends converging. It usually takes about an hour to reach this stage from the
moment when the threads become coarsely granular. The chromosomes at first sight appear practically homogeneous, but under a
higher power, $\times 1000$ for instance, slight indentations can be seen in
their surface, indicating that they are in fact formed out of a series of
discoid segments. A few minutes later the two groups, each containing
half the chromatic substance of the original nucleus, separate from each
other in the lengthwise direction, each one making for the poles of the
cell. After another five minutes they lie quite an appreciable distance
apart from each other. The daughter-chromosomes become shorter
and thicker, whilst between the two groups the crystal-clear mass of
cytoplasm keeps on increasing visibly in quantity. Within this
transparent central mass finer threads can be distinguished which
assume an arrangement like the staves of a barrel. Some twenty-five
to thirty minutes after separating out has commenced, rows of dark
dots make their appearance in the equatorial plane, and form the so-
called central cell-plate, by fusion together into a dark line which is

the membrane separating the two daughter-cells. A delicate cellulose wall is the next thing to develop. If the central barrel-like formation of cytoplasm has been wide enough to fill the whole breadth of the cell, the partition membrane between the two daughter cells will reach all parts of the side walls of the mother-cell. If, on the other hand, this formation be not of sufficient volume to stretch right across the cell's cross-section, it will tend to one side of the cell until making contact with a part of the lateral wall, to form the partition membrane of the corresponding portion in its equator. The little barrel then continues its circular motion so as to come successively into contact with the whole internal surface of the cell membrane. As this movement goes on, the dividing membrane between the two cells continues forming on the spot as previously described. In this manner the membranes are finally completed and the cells divide one from the other. The central body then withdraws slightly from the already existing diaphragm, making up the missing portions thereof out of cell-plate segments formed in the neighbourhood. Thus a cell-plate is first produced equidistant from the two halves of the nucleus, in the middle of the transparent protoplasm and from it the new diaphragm is formed.

During these changes, the daughter-chromosomes bend inwards, sending their equatorial ends towards the axis of the nucleus, ever increasingly until they make contact one with the other and become the meridian lines of a sphere. The nuclear membrane then immediately reappears and clothes each daughter-nucleus, whilst the chromosomes attenuate and lengthen, becoming finely granular and acquiring once more the original appearance of irregular windings of delicate threads on a highly complicated reel.

The coils tighten up more closely together whilst the threads increase in number and stretch out, finally anastomosing or bridging together so that the individuality of the chromosomes is lost, and their substance remains in the form of fine granules, as it was in the mother-cell before mitosis commenced.

The daughter-nuclei swell in volume, most likely at the expense of the neighbouring cytoplasm; they become cloaked in cytoplasm which extends right up to the surface, leaving vacuoles characteristic of the cell, and at the end of about an hour and a half from the start of mitosis, the separation of the daughter-cells is complete and nucleoli are plainly visible again within the nuclei.

Reproduced by permission of the publisher and Professor Koernicke

WALTHER FLEMMING 1843–1905

Zellsubstanz, Kern und Zelltheilung (Cell Matter, Nucleus and Division),
Leipzig: Vogel, 1882.

The various phases of nuclear division are clearly observable in large
cells with relatively small nucleus, such as amphibian larva tissues and
the embryo sac of plants, where the sphere of evolution of the nuclear
elements is sufficiently large for movements to be seen with less
vagueness.

We can distinguish several periods in karyokinesis or mitosis—

THE PREPARATORY PHASE

The nuclear chromatin disperses in the resting state to a distance
from any reproductive activity and, when karyokinesis is about to
commence, reassembles into a thread formed of groups of beads with
channels running through it, as observed by Balbiani in *Chironomus*.
The chromatin is distributed in stainable granules or small discs, the
microsomes, separated by colourless plastin. At the same time, within
the cytoplasm, the middle of the cell becomes divided into two
centrosomes which develop into star-shaped configurations and act
as the directive spheres. Karyokinesis now starts on its course.

THE NINE PHASES OF KARYOKINESIS

First Phase

The two directive spheres leave each other and arrange themselves
at diametrically opposite ends of the cell, their radii lengthening into
two cones coaxial on the same diameter, the apices of which are the
corresponding centrosomes whilst the bases of the cones approach
one another in order to form the protoplasmic barrel or spindle. The
central radii of these cones penetrate the nuclear membrane and thrust
the thread of chromatin towards the equatorial plane. In a very short
time the nuclear membrane is reabsorbed into the cytoplasm. This is
the "achromatic amphiaster" and "chromatic spireme" phase.

Second Phase

Development of the equator plane of the cell takes place. The
achromatic spindle does not change but the groups of nuclear beads
arrange themselves in festoons on the plane of the equator, forming
a starlike configuration. This is the "chromatic aster" phase.

Third Phase

The aster breaks up at its points into as many V- or U-shaped segments bent towards the centre of the cell as there were branches originally. These V-shaped segments are the chromosomes, and this is the "nuclear V" phase.

Fourth Phase

Each chromosome straightens out lengthways, as each particle of chromatin or microsome splits into two equal grains. This is the "twin chromosome" or "nuclear W" stage.

Fifth Phase

By double gyration around the equatorial plane, the two elements in each W are arranged symmetrically on one meridian of the achromatic spindle, their respective vertices tending towards one or other of the poles of the spindle. This is the "pole seeking" phase.

Sixth Phase

Attracted by the respective poles, the two groups of half-chromosomes shift symmetrically on their corresponding meridian. This is known as the stage of "attraction toward the poles." The spindle keeps on producing cones one after the other, each shorter than the preceding one, and between them hardly visible parallel filaments make their appearance as generating lines of a cylinder joining each two semi-chromosomes of the same pair. It has not yet been decided whether we should consider these filaments as originating from the nucleus or from the cytoplasm.

Seventh Phase

Arriving simultaneously at their respective poles, the groups of half-chromosomes group into two stars, each around the corresponding centrosome. This is the "chromatic amphiaster" stage.

Eighth Phase

The two asters at the poles change into subdivided groups. This is the "double spireme" stage.

Ninth Phase

To end up, each of the spireme groups is surrounded by a membrane thus giving place to two daughter-nuclei. This is the "twin nucleus" stage. Simultaneously the rest of the cell splits into two equal daughter-cells. In this way the sharing of chromatin from the mother-cell between the two nuclei of the daughter-cells is effected rigorously in precisely equal parts, which is the object of karyokinesis.

Reproduced by permission of the Springer-Verlag

YVES DELAGE 1854–1920
and EDGARD HEROUARD *b.* 1858

Traité de Zoologie Concrète (Treatise on Practical Zoology), Paris: Schleicher, 1896.

CELL DIVISION THEORIES

Various interesting attempts have been made in an effort to discover the causes of cell division, whether they be mechanical or determinant.

With regard to the mechanical causes two forces have been suggested, on the one hand a contraction of the spindle filaments and on the other a chemotactic attraction exerted on the chromosomes by the centrosomes. The former opinion was propounded by E. van Beneden in 1887, according to whom, the filaments shot out from the spheres of attraction become attached directly to the twin loops and drag them towards the poles whilst contracting. This tractive effect might even be the cause of the longitudinal splitting of the chromosomes. This idea has been supported by a large number of writers, including Boveri, O. Hertwig, Bergh, Rawitz, C. Schneider, and Rabl, the last of whom has given a very complete description of the manner in which these phenomena run their course.

However, Flemming in 1891 and Hermann after him raised a most powerful objection against this attractive theory, when they pointed out that the lengthwise splitting of the chromosomes often precedes that of the asters and centrosome. The theory may be partly true, but it is erroneous in certain respects and insufficient in others. Certainly the division of the centrosome and that of the chromosomes are perhaps associated without one being the direct cause or direct effect of the other.

The second alternative was advanced by Strasburger in 1893. This scientist held that the wings slide only on the filaments being attracted by some chemotactic force (or rather should we say biotactic) emanating from the attractional spheres.

Still more theoretical are the forces invoked as determinant causes of cell division.

We know that Spencer, quite a long time ago, in 1864 called attention to the fact that when an organic being grows, its surface area increases as the square of its linear dimensions and its volume as the cube. Assimilation must be proportional to volume and, as it takes place only through the surface, the consequence is that the more a creature grows the more difficult its nutrition becomes. Von Rees in 1887 applying these considerations to the cell thinks that the reasons for division lie here. But, showing that a thing is advantageous does not explain *why* it happens. It would be necessary to prove how the difficulty in nutrition became the physiological stimulus for division. Rees sought to show by certain examples that this stimulative action from lack of nourishing power is a fact; in protozoa unfavourable conditions do cause division. Orr strove to determine the source of the movement causing division in the relative asphyxia resulting from diminution of the respiratory surface as compared to volume.

Every cell, whilst living, accumulates within itself spent products which it has to expel, and usually gets rid of them by excretion. Lendl in 1890 thought this could also be achieved by the cell splitting off from itself the portion of its body where these products, which he called *ballast*, had accumulated. Division in such case would have its origin in excretion of which it would be merely one particular aspect.

Conjugation is an inverse process to cell division by which two distinct cells blend into a single one. It results in the formation of one fresh cellular individual formed from the substance of two different cells. The race thereby gains an afflux of life which is apparent in the cells issuing from divisions following conjugation, and is preserved for a longer or shorter period, gradually becoming exhausted as asexual generations succeed one another until it is renewed by a fresh fusion.

Such degeneration of the race after too long a series of divisions was excellently observed in the ciliates by Maupas, who gave it the expressive name of senile decay. Seeing that conjugation is an essential condition for indefinite reproduction by segmentation among all creatures where it does exist, we might be tempted to generalize on this and believe it to be essential to every creature without exception; but a fair number of algae and the majority of fungi are reproduced exclusively by asexual spores. There are therefore some beings amongst whom assuredly conjugation does not exist, though in many others we can only say that it has not yet been observed, and that through advances in science this number decreases every day.

CHAPTER VII

GERM-CELLS AND SOMA: SEXUAL AND ASEXUAL REPRODUCTION

DIFFERENTIATION OF GERM-CELLS

WHEN multiplying, cells may separate out, individual ones breaking away, or perhaps remaining joined together in clusters and increasing continually in volume through the division of individual component cells. In some cases this increase in the total quantity of cells goes on indefinitely. In others, however, the species has a characteristic limit, namely the colony. Transition is possible from single-celled organism to polyplastid. Such multiplication of cells may occur in one dimension, as in the filamentary systems; in two dimensions, as in membranes; or in three dimensions, i.e. volume. In the more primitive forms of coexistence, no special cells seem specifically earmarked to reproduction, i.e. there are no germ-cells as such. All the cellular elements composing them may be equally divided, and it is from such individual division that growth and reproduction of the whole is attained. This is observable among bacteria in general, in certain ascomycete fungi or yeasts, and in certain unicellular algae such as the genus *Protococcus*, etc.

As a contrast, in other algae, some cells become functionally distinct from the remainder, because their protoplasm and nuclear matter are divided into protoplasmic formations constituting spore. When the ripening thereof is completed, the cell membrane is broken and the spores stream abroad. Each spore on finding a suitable environment will give rise to the development of a fresh individual. This is a type of specialization which causes a particular cell to take upon itself the functions of reproduction, and occurs in single-celled algae like *Sphaerella* or in multicellular filamentary algae like *Ulothrix*.

REPRODUCTION BY GAMETES

However, alongside reproduction by spores, a form of sexual reproduction also develops in various species. Certain cells form protoplasts from which issue cellular elements, smaller than spores, namely the gametes. These are similarly released by fracture of the membrane of the cell where they are generated and they have to unite two by two in order to fuse and enable a zygote to be produced. From the zygote, by successive divisions, spores also are obtained. The spores, like the zygote, are capable of latent existence; they multiply only under favourable conditions to the point of producing a fresh individual.

In the same way as among rudimentary plants, among certain protozoa in colonies, there is the twofold possibility of direct reproduction and reproduction through specialized germ-cells. When these make their appearance, the other or somatic cells become more and more incapable of reproduction.

Colonies are made up of combinations of single-celled protozoa, of which a good example is the genus *Chlamydomonas*, whose cells are ovoid with two flagella, two contractile vacuoles containing a chromatophore with red pigment and another more voluminous with chlorophyll. Because chlorophyll is present and because the cell is surrounded by a thick membrane, this genus was at one time thought to belong to the vegetable kingdom. It reproduces by cellular division with separation of the individuals so formed, or by union of two similar individuals—isogametes—and immediate redivision.

Forms of this class are united in varying quantities according to genus and species. Thus, colonies of *Gonium sociale* are made up of four individuals, whilst *Gonium pectorale* contains sixteen, *Pandorina morum* eight or sixteen, *Eudorina elegans* eight, sixteen, thirty-two or sixty-four, *Pleodorina illinoisensis* thirty-two, *Pleodorina californica* sixty-four or a hundred and twenty-eight, and so on and on. In volvox types such as *Volvox globator*, the number of individuals making up a colony may run into thousands, arranged on the surface of a sphere whose interior is found to be full of watery fluid. Each of the component individuals in such a colony is a flagellate cell, with two flagella to the outside, similar to *Chlamydomonas*.

In the simpler combinations among these organizations, cellular reproduction comes about through simultaneous division of each one of the cells, that is to say symmetric reproduction whereby each colony is divided into two. It is also feasible for individuals to break off from the colony in order to unite again immediately each with an identical mate, like isogametes, forming a zygote by such union. Divisions of such zygote will generate a fresh colony. Even so, this does not yet establish specialization of germ-cells.

In other cases, in more populous colonies, some of the cells making them up are engaged in the reproduction of the colony whilst the remainder, the somatic cells, lose the reproductive faculty. In *Pleodorina illinoisensis*, of the thirty-two individuals making up the colony, twenty-eight are large cells suitable for division to form gametes, whilst the remaining four are small and sterile. In *Pleodorina californica* with sixty-four or a hundred and twenty-eight cells, approximately half are germ-cells and the remainder somatic.

In *Volvox*, the greater part of the cells are somatic and a small number becomes differentiated into germ-cells. This is the start of anisogamy, which is the interesting feature of this example. Microgametes, or sperm elements, and macrogametes, or ovular elements, are produced. From the union of the former with the latter, one of each in every case, we get the zygote generating a new colony by progressive division.

These organisms mark the transition between single-celled organisms and polyplastids. They are not yet polyplastids, because there is no morphological and functional differentiation of the somatic cells in various tissues; moreover, physiological correlation to unify and determine individuality is lacking.

Among primitive plants, certain fungi, phycomycetes for instance, the formation of gametes is also to be seen in its initial stages. Hypha cells belonging to various branches of mycelium approach one another before differentiation, eventually making contact, and then each of the touching cells behaves like a corresponding gamete. The two gametes immediately unite and form a single cell, the zygote, whence spring the spores to reproduce the plant.

Sporulation initially, like the formation of gametes, takes place in cells in no way distinguishable from their fellows, which, nevertheless, for reasons still unknown, behave like germ-cells specializing in the function of forming fresh individuals.

CONJUGATION AND FERTILIZATION

The union of gametes, the addition of two cells one to the other, or amphimyxis betokens fusion of the respective nuclear substances. This is a universal process throughout the world of life, much the most frequent in the production of simple and complex plants and animals. It is a matter of *reorganization* of the elementary unit of any organism such as a cell, and is the result of nuclear interchange.

The case of *Spirogyra* in plants is typical. Two filaments lie parallel alongside each other, each emitting an evagination starting through the membrane. These evaginations unite at the tip, perforating each other's end and forming a connecting tube perpendicular to the centre-line of the threads, thus joining the two cells. One of the two penetrates into the other, by means of the tubes, the two nuclei unite and a single cell is thus formed, the zygote, which divides in profusion to reproduce the plant.

Comparable with this is the conjugation of certain protozoa. The process is the same as that above described, though here it takes place in rudimentary animals. This was a phenomenon of great interest discovered a lifetime ago by Maupas, in 1889.

In certain *Cilia*, for instance *Paramecia*, two individuals approach one another, making contact with their cytostoma, between which a protoplasmic bridge is formed. These protozoa contain two nuclei, a large one or macronucleus which is trophic, and a small one, the micronucleus, which is the reproducer. The large nucleus breaks up and is dissolved in the protoplasm at the start of the process. Meanwhile, the small nucleus divides mitotically twice in succession, so that four micronuclei are thus formed in each of the conjugate individuals. Three of these then disappear into the cytoplasm and the remaining micronucleus lengthens out, approaching the protoplasm bridge joining the two individuals. Such lengthening-out causes the micronucleus to split into two halves, of which one stays in its own part, whilst the other passes over to the conjugate. As this passage occurs simultaneously in each of the two individuals, the result is an exchange of half the chromatin from each of the two micronuclei. Once this has taken place, the two paramecia separate, whilst the blended nucleus of each one, being the sum of two half-nuclei, splits into two portions, whereof one hardly grows at all, to become the micronucleus, whilst the other part increases in volume to form a new macronucleus.

It was originally thought that the exchange of nuclear chromatin between two individuals would intrinsically bring about the "rejuvenation" of each, and give them new life, being thus the manifestation of a law according to which germ plasms would enjoy everlasting life. The fashion was to talk of the "immortality" of infusoria, which immortality has, however, not been confirmed by facts. Conjugating infusoria also grow old and die just like anything else.

SPORES AND GAMETES

Germinal specialization can likewise lead to direct sporulation without previous union of the gametic cells, and to reproduction by cellular conjugation. This can, on the other hand, be brought about by union of like cells or *isogamy*, or even by different gametes large and small, male and female, which is *anisogamy*. In the latter case, the micro-gamete is, as its name implies, small and mobile and seeks out the macrogamete until it can introduce its nuclear substance into the latter. The macrogamete usually contains in its cytoplasm reserves of nutrition to support the metabolic needs for development of the embryo which make themselves felt when cellular divisions of the zygote commence.

Reproduction by gametes appears in very simple forms among living bodies and is usually isogamic under such circumstances, as in the monocellular algae, such as *Sphaerella* or in the filamentary *Ulothrix*, or it may be anisogamic as in *Oedogonium* and other cases, and also in rudimentary animals. Thus, in the hydroids, the production of macrogametes and microgametes, ovaries and testes, is observable. Temporary conical projections develop in the ectoderm formed by embryonic interstitial cells. Sperm elements sally forth herefrom dropping freely into the fluid and fertilizing the ovule adherent to the ovary of the corresponding individual. The fertilized ovule, the zygote, then loosens off, and becoming progressively divided gives place to a new hydra.

The possibility of forming spores occurs now and then in cells which are indistinguishable from the rest, but shows itself generally in cells forming part of special organs, as in the sporangia. Similarly, gametes originate in gametangia: antheridia, whence proceed the micro-gametes and archegonia, the formation foci of the macrogametes.

ALTERNATION OF GENERATIONS

In many cases reproduction through direct sporulation alternates with that through union of gametes. It is to be seen in unicellular animals and plants, various protozoa and algae, and particularly when, at distinct stages in their existence, they inhabit different media.

This occurs in various types of living organisms. Thus, in species of mosses and in the ferns, which exhibit generations of gametophytes interspersed among generations of sporophytes. Among the same higher plants, the phanerogams, we can distinguish an asexual genera-tion followed by a sexuate one in uninterrupted and continuous alternation one after the other. Numerous examples are to be found, particularly among protozoa; one well known case is that of the malarial parasite *plasmodium*, where we get alternating sexuate, gametic, amphiontic reproduction of haemosporidia in the transmittor, the anopheles mosquito, and sporular, agamic or monontic reproduc-tion thereof in man.

Similar instances are to be found in certain polyplastids and the alternation of generations in hydrozoa is a typical case. The colony is but one phase in the sequence of generations in this creature. From it, at some given moment, through development of a yolk, a free individual breaks loose, the medusa, capable of independent existence on its own in the sea. The colony is reproduced asexually and the medusa sexually. The hydra forms a similar colony in which the medusa form does not make its appearance. In other coelenterata, the scyphomedusae, the alternating process is likewise seen at work giving asexual generation by strobilation in the fixed forms, and sexuate in the mobile forms of the medusae.

Animal species do exist with germ tissue or gonads which appear temporarily at particular seasons of the year or by reason of other special circumstances. In other species, the majority in fact, the sexual organs are permanent.

DIFFERENTIATION IN GERM-CELLS

The germ cells characterizing these organs come about from a morphologic and functional differentiation like any other, and identical to that forming the source of the various tissues in embryonic development. In some creatures, as for instance worms of the genus *Ascaris*, or *Sphaerium* among the molluscs, it is possible to distinguish germ-cells from the somatics, from which the organs arise, right from the commencement of blastomeric division. In general, however, the distinction between one and the other has to be made at a later stage. Primitive germ-cells form localized areas in the epithelium of the coelum, or on one side or the other of the spine in vertebrates. The reproductive cells develop there in conjunction with connective tissue, blood vessels and nerves. Differentiation may take place much later, when the adult stage has already been reached, even in the higher animal forms, such as the mammifera themselves. Davenport in 1925, Parkes, Fielding and Brambell in 1927, state that in mice, regeneration of the ovary may be achieved after its having been taken out, by formation of fresh germ-cells at the expense of the somatic cells.

Similarly in plants, we find that the meristems, from which branches, leaves, and flowers derive, differentiate germ-cells from somatic, without the former being distinguishable from the others by any special cytological characters even from their first moment. Once differentiation is established, the germ-cells become localized in the corresponding

organs and emigration of these germ-cells in the individual then does
not appear to be possible.

GERM PLASM

Following Nussbaum, Weismann in 1885 propounded his theory of
the "continuity of the germ plasm." According to this, germ-cells
pass from generation to generation, apparently immortal, in contrast to
the somatic cells which make up the individuals, and consequently die
off. The biological part played by each individual, with its somatic
cells, seems to be merely as a mortal support for the germ-cells, in their
uninterrupted course through the ages. S. Butler comments: "A hen
is apparently nothing more than a means contrived by the egg so that
another egg may be hatched."

By considering the germ cells as quite apart from the somatic cells,
Weismann came to the conclusion that there was no connection
between the one and the other within an organism, but a complete
separation, an absolute partition between them, such that the somatic
cells would not and could not exercise any influence whatsoever upon
the germ-cells, whereby there could be no possibility of hereditary
transmission of characters acquired by an individual in the course of its
life. This would be worthy of discussion if germ-cells did indeed
remain in all cases demarcated in the first division of the zygote.

This is the plan which was mapped out by Walter. It is not however
proven that one of the first pair of blastomeres already shows the
characters and functions of a germ-cell. There does exist, on the other
hand, the possibility, previously pointed out, of reproduction by means
of somatic cells, through which differentiation they will form germ-
cells in fresh individuals. This would obviously interrupt the germ line
and there would be no continuity in the germ plasm nor immortal cells.

The intrinsic characters of cells charged with the business of reproduction make their appearance at different moments according to the
species under consideration. In some of them, the germ-cells are very
speedily distinguishable from the somatics; in others this is not possible
until later. The germinative quality is the result of one of many
differentiations in ontogenesis; cells which acquire special reproductive
faculties have to be distinguished through their specialization to carry
out genetic functions and in the particular morphology of each one of
these classes of cells. Germinal capacity demands only the existence of
a genetic inheritance enabling each one of the characters distinguishing
a species to make its appearance in due course.

REPRODUCTION BY SOMATIC CELLS

The said reproductive capacity is not, moreover, exclusive to germ-cells and it varies even among somatic cells from one class to the other. As far as we know, there is a possibility of reproduction by somatic cells. This faculty shows itself by degrees, being absolute in the zygote and partial in a blastomere existing with one or more of its kind. If however such blastomere becomes isolated from the others and sets forth from one of the so-called regulation eggs, belonging to certain animal species, that partial faculty may become total and the blastomere on dividing will give birth to an embryo, just as would a zygote. This is of course easier the younger the generation to which the blastomere belongs.

The further we progress with embryonic development, the less becomes the range of morphogenic capacity in each cell. The sum total of the ontogenic faculty decreases progressively with differentiation, forming the various tissues. Only the germ-cells then preserve such total formative capacity.

According to what this capacity may be in its actual or its potential state, the possibility of regeneration of injuries will be more or less efficient. Among plants, this faculty is both wide and deep. In certain animals, particularly the lower ones, it is also present. It is to be seen to a more limited degree in other cases whilst in some it becomes reduced to mere cicatrization of a wound. The greater the differentiation, or what is the same thing, the higher the creature in the scale of life, the less are the possibilities of what we may call automatic regeneration. The differentiatory morphogenic power decreases in exact proportion to the degree of differentiation in the cell. The nerve cell has lost the power of division. Thus differentiation means specialization. For this reason, in the superior forms of living matter, among the animals, the reproductive function remains reserved to the germ-cells, and in plants also reproduction by seeds rather than growth from shoots is generally more certain in its effects. Reproduction seems to be a specialized function, of use particularly to animate beings of greater morphological and physiological delicacy; it is one improvement, amongst many, of biological progress through differentiation. All of which does not, however, prevent reproduction from being simply one manifestation among others of growth and cellular multiplication, first through somatic cells and afterwards through germ-cells.

* * * * *

AUGUST WEISMANN 1834–1914

Vorträge über Descendenztheorie—Die Keimplasmatheorie (Lectures on the
Theory of Descent—The Germ Plasm Theory), Jena: Fischer, 1902.

If I may now go so far as to propound a theory of heredity in the
manner it has unrolled before me in the course of my own scientific
development, I might start off by showing you that the factors, not
merely from *one single individual* but rather from *several* and indeed
often from *a great many*, are containable in the hereditary substance of
the germ-cell of the animal or plant.

I start off with what I believe is the proven fact that the hereditary
material resides in the chromatic substance of the nucleus. We have
seen that this reaches each type of germ-cell in the form of a definite
number of chromosomes, and that this number is previously halved in
the germ-cells concerned with fertilization, which is to say in the sexual
cells, and in fact as has been demonstrated for a whole range of creatures,
this is effected in the last two cell divisions, the so-called ripening
divisions.

We know that the whole number is again made up by the process of
amphimyxis, since the half-quantities of chromosomes from the male
and female germ-cells combine in a *single* cell, the "fertilized ovum,"
and in a *single* nucleus, the so-called *fission nucleus*. The hereditary
material of the offspring is thus formed half from the paternal and half
from the maternal substance of transmission, and we have seen that this
remains true during the whole development of the offspring, since
in each subsequent cell-division each of the paternal and each of the
maternal chromosomes is doubled by partition and the resultant
halves are distributed to the two daughter-nuclei.

If now the whole hereditary material of a germ cell *prior* to the
reduction division contains latently the whole of the elements of the
body, as is obvious, then after the reduction each germ cell must contain
either only half the components from the parents—or *the whole
number of elements must be contained collectively in the half-quantity of
chromosomes*. The latter alternative seems to me to be the only logical
supposition, as I shall soon show, and by that it is clear that *at the very
least, factors for two complete individuals must be contained in the chromosomes of the fission nucleus*.

That this conclusion is justified follows indeed from the fact that an entire, that is to say a complete individual, develops with all its parts from the egg, but a damaged one does not. Supposing, therefore, that each mature germ-cell contains only half the bodily factors, then it would be impossible for such halves, as they would be distributed by the luck of the division, to complement each other exactly in the two cells uniting in fertilization and it would therefore be much more likely that they were *not* properly paired, so that from their union an individual might result lacking in certain parts. From the combination of the two sexual cells, we should scarcely ever get an embryo complete with all its parts, for sometimes one and sometimes another set of parts would necessarily be lacking whilst some other might be doubled in size or be formed in duplicate.

Moreover, the facts of heredity teach us that resemblance to female and male parent can occur simultaneously in *all* or even in *particular* parts of the offspring, as is especially evident in plant hybrids, whence it inevitably follows *that all constituents of the entire organism are in fact present in the half-quantities of chromosomes.*

From now on I shall call the inheritable substance of a cell its *idioplasm*, following the example of Nägeli, who in fact expected it in the *body* of the cell, not in the nucleus, and theoretically thought it to be different in action, though as we have seen he was the first to conceive and formulate the *idea* of it as an *elemental substance* determining the entire structure of the organism in contrast to the ordinary protoplasm. Every cell contains idioplasm, but since each also contains chromatin in its nucleus, I denote the idioplasm of the germ-cell as *germ plasm*, the elemental material for the entire organism, whilst the *complexes* shown to be equally available for the requisite elements of an entire individual I term *ids*. These ids may in many cases coincide with the "chromosomes" at least in all those where the chromosomes are of *simple* composition, that is, not made up of several congruent formations. Thus, in the case of the brine shrimp *Artemia salina*, with its 168 small granular chromosomes, each of these chromosomes must be considered as an id, since every one of them can under certain circumstances be driven from the egg at the reduction division and combined with a great variety of other chromosomes on fertilization. Each of them must therefore be complete germ plasm in the sense that all parts of an individual are virtually contained therein: *each one is a biological unit— an id*. If, however, in many animals we observe sizable interlooped or even rod-like "chromosomes," and if these, as for instance in the oft referred to *Ascaris megalocephala*, are composed of a series of granules, then each of these granules may be looked on as an id. Indeed, we then discover in other forms of Ascaris a larger number of small globular chromosomes instead of the two or four large rod-shaped chromosomes appearing in *Ascaris megalocephala*.

Combined chromosomes consisting of several ids, as in fact all rod-shaped or loop-shaped elements of nuclear matter are, I term *idants*. Because of their smallness, their composition from several single ids does not always clearly stand out, and even in the larger of them only at certain stages. In the salamander sperm-cell the single ids are not visible at the earlier stage, though later on when the loops split a necklace formation becomes evident. We cannot therefore regard each chromosome immediately as corresponding to *one only* or *several* ids. On a more precise delving into the processes of the reduction division, it has been shown that there are *multivalent* chromosomes made up, that is, of *several* ids, though their plurivalence is not directly recognizable but can only be deduced from their later development; there are bivalent and quadrivalent chromosomes which we must consider as made up out of two or four ids respectively. It would lead us too far afield if I were to go further into this matter here, and the plan and purpose of this lecture does not oblige us to go deeper into this most intimate and even now still much debated question.

The germ plasm of each kind of animal and plant is therefore made up of a greater or less number of ids or individual elements, and only through their interaction does the individual developing from the egg become determined.

Now at last we have reached the point at which it is most fitting to discuss the *organization of living matter* in general.

The Viennese physiologist, Ernst Brücke, some forty years ago established the idea that living matter might be not merely a conglomeration of all sorts of chemical particles, but could well be *organized*, which is to say made up of minute invisible vital units. If —as we must therefore suppose—the *mechanist* theory of life is correct, if there is no vital force in the natural philosopher's meaning of the term, then Brücke's hypothesis is incontestable, for a chance mixture of molecules cannot produce the vital phenomena, not even a *single* molecule, since mere molecules do not live empirically, neither do they assimilate nor do they grow or propagate themselves. Life can therefore only arise through a definite combination of various types of molecule, and all animate matter must consist of such definite groups of molecules. Herbert Spencer a little while later than Brücke also assumed such vital "units," and still more recently de Vries, Wiesner and I myself have adopted this idea. On the composition of these vitality carriers, or *biophors* as I call them, we cannot for the time being say anything more precise than that the protein molecules, water, salts and some other substances play the main part in their make-up. This follows from the chemical analysis of dead protoplasm; the form in which these materials are contained in the biophor and the way they act on one another in order to call forth the vital phenomena whilst passing through a continuous cycle from decomposition to reconstitution is, however, still quite hidden from us.

We do not need to bother about that here; it is sufficient to attribute the character of life to the biophors, and to consider all living substances, therefore, cell and nuclear matter, muscle, nerve and gland materials in all their variants, to consist of biophors, which latter must naturally be most varied in their composition. There must be countless types of biophors, in all the different parts of the millions of forms of life which now exist on our planet; all of them, however, in order to exist must be built up according to some particular basic plan which at the same time determines their remarkable activity; all of them possess the fundamental property of animation, and disassimilate, assimilate, grow and multiply by partition. We must also grant them movement and sensation to some degree and intent.

As to their dimensions, it can only be said that they lie far below the limits of visibility and that all the smallest granules of the protoplasm, as far as they are directly ascertainable by means of our most powerful instruments, cannot be single biophors but masses thereof. On the other hand, however, they must be larger than any chemical molecule whatsoever, since in fact they themselves consist of a group of molecules, among which they are to be found in the most complex formation and must likewise be of corresponding relatively significant size.

It may then be asked whether perchance the previously inferred determinants are not identical with these "biophors" or infinitely small vital particles; this is not *generally* the case, however. We gave the name of determinants to those particles of the embryonic matter which determine a "hereditary fragment" of the body, i.e. from whose presence in the embryo it depends that a fixed part of the body whether composed of a group of cells, of one cell only, or of a part of a cell, shall be formed and that in a particular manner, and from whose variation furthermore only these particular parts can be induced to vary correspondingly.

We now ask how large and how numerous such hereditary portions are, whether each cell or each part of a cell in the body represents one such fragment, or whether only larger groups of cells play this part. It is now evident that these agglomerations, *uniquely* definable from the very germ, are quite different in size one from the other according to whether we are dealing with small or large, simple or complex organisms. The single-celled creatures, for instance infusoria, must indeed possess special determinants for a medley of cellular organs and parts, if we are unable to identify directly even the independent and inheritable variability of their organs; lower multicellular organisms, such as perhaps the chalky sponges, will need only a relatively small number of determinants, but among the higher multicellulars, e.g. certainly in the majority of articulated animals, their number must already be extremely high and amount indeed to several thousand if not to hundreds of thousands; for here everything in the body is already specialized and must be modified by autonomous variation right from the embryo itself.

Reproduced by permission of the publisher

HENRY FAIRFIELD OSBORN 1857–1935

The Origin and Evolution of Life, New York: Scribner, 1917.

EVOLUTION OF PROTOPLASM AND CHROMATIN, THE TWO STRUCTURAL COMPONENTS OF THE LIVING WORLD

The development of the cell theory after its enunciation in 1838 by Schleiden and Schwann followed first the differentiation of protoplasmic structure in the cellular tissues (histology). Since 1880 it has taken a new direction in investigating the *chemical and functional separation of the chromatin.* As protoplasm is now known to be the *expression*, so chromatin is now known to be the *seat* of heredity, which Nägeli (1884) was the first to discuss as having a physico-chemical basis; the "idioplasm" postulated in his theory being realized in the actual structure of the chromatin as developed in the researches of Hertwig, Strasburger, Kölliker, and Weismann, who independently and almost simultaneously (1884, 1885) were led to the conclusion that the nucleus of the cell contains the physical basis of inheritance and that the chromatin is its essential constituent. In the development from unicellular (Protozoa) into multicellular (Metazoa) organisms, the chromatin is distributed through the nuclei to all the cells of the body, but Boveri has demonstrated that all the body-cells lose a portion of their chromatin and only the germ-cells retain the entire ancestral heritage.

Chemically, the most characteristic peculiarity of chromatin, as contrasted with protoplasm, is its phosphorus content. It is also distinguished by a strong affinity for certain stains which cause its scattered or collected particles to appear intensely dark. Nuclein, which is probably identical with chromatin, is a complex albuminoid substance rich in phosphorus. The chemical, or molecular and atomic, constitution of chromatin infinitely exceeds in complexity that of any other form of matter or energy known. As intimated above, it not improbably contains undetected chemical elements. Experiments made by Oskar, Gunther, and Paula Hertwig (1911–14) resulted in the conclusion that in cells exposed to radium rays, the seat of injury is chiefly, if not exclusively, in the chromatin; these experiments point also to the separate and distinct chemical constitution of the chromatin.

The principle formulated by Cuvier that the distinctive property of life is the maintenance of the individual specific form throughout the incessant changes of matter which occur in the inflow and outflow of energy, acquires wider scope in the law of the continuity of the germ-plasm (i.e. chromatin) announced by Weismann in 1883, for it is in the heredity-chromatin that the ideal form is not only preserved but, through subdivision, carried into the germ-cells of all the present and succeeding generations.

It would appear, according to this interpretation, that the continuity of life since it first appeared in Archaeozoic time is the continuity of the physico-chemical energies of the chromatin; the development of the individual life is an unfolding of the energies taken within the body under the directing agency of the chromatin; and the evolution of life is essentially the evolution of the chromatin energies. . . .

We are equally ignorant as to how the chromatin responds to the actions, reactions, and interactions of the body cells, of the life environment, and of the physical environment, so as to call forth a new adaptive character, unless it be through some infinitely complex system of chemical messengers and other catalytic agencies. Yet in pursuing the history of the evolution of life upon the earth we may constantly keep before us our fundamental biologic law that the causes of evolution are to be sought within four complexes of energies, which are partly visible and partly invisible.

Reproduced by permission of the publisher and G. Bell & Sons, Ltd., London

JULIAN SORELL HUXLEY *b.* 1887

Essays of a Biologist, London: Chatto and Windus, 1923.

SEX BIOLOGY AND SEX PSYCHOLOGY

In the first place, then, we have to consider the evolutionary history of sex. Of its origin, we can say only that it is veiled in complete obscurity. Once present, however, it appears to have a definite function by making possible, through sexual reproduction, all the various combinations of any hereditable variations that may arise in different individuals of a species, and so conferring greater evolutionary plasticity on the species as a whole. (*See* East and Jones, *Inbreeding and Outbreeding*, Philadelphia, 1919.)

Primarily, sex implies only the fusion of nuclei from two separate individuals; there is no need for sex differences to exist at all. Sex differences, however, are almost universal in sexually-reproducing organisms, and represent a division of labour between the active male cell and the passive female cell, the former taking over the task of uniting the two, the latter storing up nutriment for the new individual that will result from that union.

The subsequent history of sex is, roughly speaking, the history of its invasion of more and more of the organization of its possessors. First the male as a whole, and not merely its reproductive cells, tends to become organized for finding the female. The female's whole type of metabolism is altered to produce the most efficient storage of reserve material in her ova, and later she almost invariably protects and nourishes the young during the first part of their development, either within or without her own body. Appropriate instincts are of course developed in both male and female.

At the outset there is enormous waste incurred in the liberation of sperms and ova into the water, there to unite as best they may. Congress of sexes eliminates the major part of this waste, and is universal above a certain level. This is in itself the basis for other changes. As the mind, or shall we say the psycho-neural organization, becomes more complex, the sexual instinct becomes more interwoven with the general emotional state; and a large number of animals appear not to mate unless their emotional state has been raised to a certain level. The result of this is that special actions, associated generally with bright colours or striking structures, with song or with scent, come into being.

The exact mechanism of the appearance of these courtship-displays is a much-vexed point; but it is undoubted that they only occur in animals with congress of the sexes with minds above a certain level of complexity, and that they are employed in ceremonies between the two sexes at mating-time. There can subsist no reasonable doubt that there exists some causal connection between the associated facts.

An important point, which has been commonly overlooked, is that such characters and actions may be either developed in one sex only, or in both. In a large number of birds, such as egrets, grebes, cranes and many others, the courtship-displays are mutual, and the characters used in them developed to a similar extent in both sexes. Such characters are therefore often not secondary sexual differences, and we had best use Poulton's term *epigamic* for them, whether they are developed in one or in both sexes. . . .

In vertebrates the gonads form part of what has been called the chemical directorate of the body—the interlocking system of endocrine glands, each of which is exerting an effect upon the rest. It is only in higher groups that these emotion-stimulating sexual characters arise, for only in them has mind reached a sufficient degree of perfection.

When we reach man, however, the whole aspect of the matter changes. The change is most marked, naturally, in his mental organization. Through his powers of rapid and unlimited association, any one part of his experience can be combined with any other; through his powers of generalizing and of giving names to things, his experience is far more highly organized than that of any animal; through speech and writing, he is inheritor of a continuous tradition which enormously enlarges his range of experience. Again, he can frame a purpose and thus put the objective of his actions far further into the future than can lower organisms.

There are, however, also changes of considerable biological importance on the physical side. Man brings with him from his animal ancestors the endocrine secretory mechanism of the reproductive organs: but his life is not subordinated to it in such an iron-bound way. This is not to deny that the sexual life of man is dependent upon the reproductive hormones. It is apparently necessary for proper activation of the sexual centres of the brain that there should occur a continuous liberation of secretion from the reproductive organs into the blood. Again, the mental activities of man are so much more important than those of other forms, that even the cessation of activity of the reproductive organs, for instance, in the female at the change of life, or even their total removal, need not prevent the continuation, albeit in a modified form, of the sexual life in its varied indirect manifestations.

It is quite clear from these and other facts that in higher vertebrates there are present in every individual of either sex the nervous connections which give the possibility of either male or female behaviour.

Reproduced by permission of the publisher and Alfred A. Knopf Inc., New York

CHAPTER VIII

FORM AND DYNAMICS OF REPRODUCTION

THE FERTILIZATION PROCESS

WHENEVER a cell divides either directly or by mitosis, the most striking thing is the sequence of events developing in the nuclear substance. The discovery of the various phases of karyokinesis was very soon followed by that of the morphology of fertilization or *syngamy*. Penetration of the nucleus by the male gamete or pronucleus is the natural way of exciting mitosis in the ovule through the ensuing action upon its nucleus, that is to say on the female pronucleus. The chromatic substance of both nuclei contributes on an equal footing to karyokinesis, from which division of the fertilized egg or zygote into two blastomeres comes about. This was observed by O. Hertwig as far back as 1875 and confirmed by van Beneden in 1883 and by Boveri in 1886 using material of a different nature and origin.

In 1865 Nägeli had assumed that germ-cells contained directional micelles. Kölliker (1875–85) held that the nucleus itself might be the carrier of hereditable properties, on the assumption that the "idioplasm" resided therein, and it remained only for the protoplasm to fulfil a nutritive function. Strasburger in 1876 describes the chromosomes, whilst van Beneden in 1883 and Rabl in 1885 indicate their significance. Weismann in 1883 stated that they are the bearers of characters transmissible through heredity. Of late the conviction has grown more and more pronounced that chromosomes play a predominant and often exclusive part in heredity. We may mention, for instance, the voluminous researches of Edmund Beecher Wilson, from 1903 onwards, with regard to chromosomes over an extremely varied range of species and their functions.

In fertilization, chromosomes from each of the two participating gametes draw up alongside each other and form half and half the branches of the mother aster or equatorial plate at the commencement of metaphase. When immediately the hitherto single aster starts to divide into two daughter-asters, it is obvious that these latter will likewise be made up, as to one half, of chromatin of paternal origin, whilst the other half will derive from the maternal source. Similarly, therefore, the nuclei of the two blastomeres are produced from such division, and afterwards likewise all cells generated during successive divisions.

The consequence of this will be that all cells ultimately derivable from the zygote will necessarily possess the same number of chromosomes, as stated by Boveri in 1886, which assertion after being the object of considerable debate has now been repeatedly corroborated. The number of chromosomes in any cell, be it body-cell or germ-cell, is fixed and definite for each particular species.

CHROMATIC REDUCTION OR MEIOSIS

This constancy in the number of chromosomes necessarily implies nullification by destruction or expulsion of half the chromosomes from each gamete. Thus, when the chromosomes from two gametes unite, the zygote will come out with the same quantity as all the other cells, as will also the blastomeres and their descendants and, consequently, by the same token, all the cells of the new organism. Fol in 1876 noted the reduction in chromosomes on the ripening of germ-cells, and this was confirmed by Strasburger in 1894, by R. Hertwig in 1906, and by many others subsequently. In 1903, Farmer and Moore suggested the term *meiosis* to denote this type of process.

The chromosomes in the cells are arranged in pairs of homologous chromosomes or alleles, of which each pair is composed of one chromosome from the male source and a second from the female, resulting in *diploid* cells containing $2n$ chromosomes. When the germ-cells ripen, half the chromosomes are lost through meiosis, and cells become *haploid* with n chromosomes.

The process whereby meiosis takes place is typical in spermatogonium. Spermatogonia derive from an original germ-cell through division, all being diploid cells. On division taking place in the spermatogonia, constituting the primary spermatocytes, chromosomes from each pair adhere to each other lengthwise by synapsis thus: 1-I, 2-II, 3-III . . . , and consequent to this, each chromosome of each pair will split also in the lengthwise direction, thus giving as a result a bundle of four chromosomes, two male and two female thus: [1-I, 1'-I'] [2-II, 2'-II'] [3-III, 3'-III'] and so on, forming *tetrads*. The primary spermatocyte divides into two secondary spermatocytes. In this mitosis, each tetrad forms a branch of the mother aster, and on these branches separating in the anaphase, the resultant chromosomes will be made up naturally of two elements alongside each other: *dyads*, perhaps 1–1′, II–II′, III–III′ . . . or maybe I–I′, 2–2′, 3–3′, and so on.

Thus the secondary spermatocyte will have the same number of chromosomes as the spermatogonium, though they will now be in joined pairs so that their quantity will seem to be only half of what it really is. Moreover, in the division of the tetrads, each of the resultant dyads will be made up of chromosomes two by two from the same line, that is, each pair will come from chromosomes from the male source or from the female source exclusively, which means that the whole of the chromosomes will be distributed between two secondary spermatocytes in such a way that the male and the female chromatin will complement each other, since the dyad that has a particular class of chromosome originating from the male will not contain any of the same class coming from the female. That is to say that, just as pairs in the germ cells are formed from one chromosome of male origin and another of female origin, here each pair will be solely male or solely female. From this it comes about that when the secondary spermatocytes split into halves—spermatids—the dyads separate from each other, the nucleus of each of these spermatids attracting the chromosomes which formed the dyads, perhaps 1–II–III and 1′–II′–III′ or I–2–3 and I′–2′–3′ in the scheme already described. Both spermatids, originating in any particular secondary spermatocyte, will contain half the chromosomes that are in the primary spermatocyte and will be haploid cells, their chromosomes being, as is obvious, those of the corresponding dyads. Through such a series of operations, there will have been generated from the primary spermatocyte four spermatids, each pair having had its source in one of the two secondary spermatocytes. Two of these spermatids—*A*—will be made up of similar chromosomes, and the other two—*B*—will be offspring from the other secondary spermatocyte and likewise identical in chromosomic content. It is easy to see, therefore, that the spermatid pair *A* resulting from one of the two spermatocytes may be simultaneously complementary, through their chromosomes, to the two spermatids *B* descending from the other spermatocyte. That is, since these spermatids possess half the chromosomes from the spermatogonia, the male and the female chromosomes will be complementary pair by pair—male-female-female and female-male-male as an example —in such manner that the sum of the chromosomes of these two types of spermatid will exactly equal the chromosomic inheritance—male-female - male-female - male-female—of the original spermatogonium.

The nuclear substance, the sperm head arising from the ripening of the spermatid, will of course have the same chromosomes as the spermatid.

Identical phenomena take place during the development of the ovule. When the oogonia divide into primary oocytes, the chromosomes group together and divide into tetrads. The formation of the second

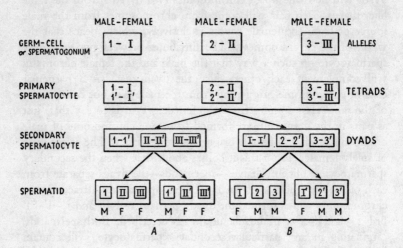

oocyte with emission of the primary cell or polar sac takes place when the tetrads split into dyads, one of which goes to the oocyte and the other, not identical, to the polar cell. The immediate division of the secondary oocyte into a ripe ovule and second polar body comes about from lengthwise partition of the dyads. One consequence of this is that the ripe ovule and the second polar body will have like chromosomes and these will be complements of those found in the nuclei of the cells resulting from the division of the first polar sac.

The process of chromatic reduction comes about therefore in the same manner in oogenesis as in spermatogenesis. The only distinguishing feature is, that in spermatogenesis the four spermatids and consequently the four derived spermatozoids from successive divisions of the primary spermatocyte are functionally complete and suitable, whilst in the formation of the ovule, one only of the four cells originating from the primary oocyte will ripen and be utilizable. The remainder of the cells, offspring from the first and second polar bodies, will degenerate and vanish.

CONSTITUTION OF THE ZYGOTE

Fresh pairs of chromosomes arise in the zygote through fertilization and the cell becomes diploid with 2*n* chromosomes, due to the super-position of the *n* chromosomes from each of the gametes.

As can be seen, in meiosis there is no question of a simple reduction of chromosomes, but of something deeper and more significant, a "segregation" or "disjunction" of chromosomes, a sharing of chromosomes as definite biological entities from one or other source, the male or the female, in such a way that the original allelomorphic pairs from the germ-cells as well as from the somatic or body cells become irregularly distributed between the two pairs of spermatids, the consequent spermatozoids, and, on the female side, between the ovule and the daughter-cells of the first polar body.

In all types of sexual reproduction, in higher and lower animals and plants, chromatic reduction is observable in the series of germ-cells giving rise to gametes of one sex or the other. Some protozoa simply expel or destroy a certain amount of chromatin but this is the exception rather than the rule; wherever it is possible to determine and count the chromosomes, it can be verified that the reduction comes about more or less in accordance with the scheme previously outlined. It never comes about by mere chance. The ripe gametes have their definite heritage of chromatin containing the portion belonging to them from the bearers of hereditary factors from one or the other line of the parent stem. Thus it is that in the make-up of the zygote, the paternal or maternal chromosomes reproduce part of the chromosomes of the four grandparents and further up the ancestral line, throughout all generations.

These combinations of chromosomes will be the more plentiful the greater the number of chromosomes from each species. This may be calculated by the mathematical laws of permutations and combinations, with however, one limitation—each gamete must receive a derivate from one of the members of each pair in the germ-cell, so that the gamete will contain one complete haploid group. It will be very difficult for such group to be formed exclusively by paternal or by maternal chromosomes alone; chromosomes from each of the two main stocks alternate one with the other in each group. In the human race, with forty-eight chromosomes in every cell, forming therefore twenty-four pairs, we can get 2^{24}, or pretty nearly seventeen million (16,777,396) types of gametes for each sex, which, through union of the gametes in fertilization means getting on well towards three hundred billion (281,481,016,540,816) chromosomic combinations possible for each pair. It is not to be wondered at, then, that different children of the same parents are not necessarily alike in every respect!

Fertilization reunites the chromosomes from two haploid gametes into a diploid zygote. It can be understood, therefore, that in the zygote, chromosomes of the same type will be duplicated from father and mother and will group up in pairs. As the chromosomes divide along their length, when the daughter-asters are generated by division of the matrix, during successive mitoses, not only during development of the embryo but throughout the whole existence of the individual, the cells of the body without exception, that is to say diploids, body cells and germ-cells alike, will always have the same number of pairs of chromosomes.

CHROMOSOMES AND GENES

The behaviour of the chromosomes in cellular division and the relation this has to the transmission of hereditable characters from one line or the other, paternal or maternal, has resulted in the recurrence of ancient theories which attributed heredity to the action of certain material particles contained in the nuclei of the germ-cells.

Nägeli's "micellar" theory (1865), Spencer's "physiological units" (1863), Darwin's "gemmules" (1868) and de Vries's "pangenes" (1878), finally culminated in Weismann's theory (1883), according to which inherited characteristics were held to be due to the intervention of material elements combined in the *idioplasm* forming part of the nuclear substance. Such idioplasm or *germ plasm* comprises echoes from every part of the organism, or *ids*. In certain cases, it will be possible to identify the ids with small chromosomes; in others, the chromosomes will be *idants* or combinations of various ids. At the same time, such ids may be complex systems made up of numerous *determinants* or organizations of minute, invisible, biological units, the living molecules or *biophors*. Originally, Weismann considered that the material unit or id might be the chromosome itself, or small chromosomes, especially when such chromosomes occur in great number in the corresponding cells. Later on, however, on taking into consideration that the chromosomes are often small in number whilst the characters which have to be accounted for are on the other hand multitudinous, he assumed that each chromosome consisted of many ids. Every id must, moreover, contain, in addition to the potential source of the character or characters it represents, the similar possibility of a source of the whole combination of characters, i.e. of the totality of characters defining the product. Each id will therefore possess a special representative element and complementary thereto a complete representation of all the characters of the individual or of the species. In guiding the course of development, the ids play their part, each one on its own account, and, on a higher level, organization of the ids will also come into the picture. Each id will therefore be a sort of microcosm, a little world in itself. Individual variations will be due to combinations of ids from the male line combining with those coming from the female source. By progressive division of cells constituting the development of the embryo, the ids keep on dividing in proportion to the multiplication of the cells, until each type of cell forming the body organs receives a *determinant*, the final result of such divisions of the ids. On the other hand, in the generation of germ-cells, there will be no disintegration of ids.

Wilhelm Ludwig Johannsen in 1909 asssumed that chromosomes are merely systems of genuine hereditary elements, and suggested the name *genes* as a label for these latter. Thomas Hunt Morgan in 1926 published his celebrated work *The Theory of the Gene*, which forms a conspectus of results obtained by the noted school of genetics which revolved around him under his direction.

The genes, distributed in exact location in the chromosome, thus facilitating the preparation of the gene charts which we now possess thanks to Mackensen's work in 1935 and to Bridges (1935-8), are present in large numbers and each one of them is responsible for the blossoming of some particular character or set of characters. Alteration of a gene is bound up with some modification of the corresponding character; the disappearance of a gene is accompanied by the disappearance of the corresponding character. On the other hand, there is withal an interrelation or mutual influence between different genes in various special rigid relationships. From the properties of each gene on its own account and from those arising from its location and proximity to other genes, we get the embryonic production of living forms and the full development and functions thereof.

CYTOPLASM AND REPRODUCTION

All these facts which we have just been reviewing corroborate the part played by the nucleus in reproduction. The question still arises, however, whether nuclear intervention is the sole factor, or whether the cytoplasm does not have some share in transmission of characters from cell to cell.

To such queries the results of experiments in *merogony*, or the fertilization of fragments of ovule by sperm penetration, give an adequate answer. Such fertilization was achieved by Boveri in 1889, when he broke up sea-urchin egg-cells by shaking. Break-up of an egg-cell is nowadays effected with precision under the compound dissection microscope. These researches have reached their present extent and importance through the work of Yves Delage (1895) in the first place, and the later investigations by Edmund Beecher Wilson (1899), von Ubisch (1925-9), S. Hörstadius (1928-35), Baltzer (1936), Curry (1936), and many more.

If fertilization is effected on a fragment of egg-cell containing its nucleus, the zygote will be diploid; but if it take place in a piece of cytoplasm denuded of nuclear matter, the resultant cell will be haploid and the maternal nuclear matter will not be able to exert any influence on the characters of the product, and indeed the ovular cytoplasm only will act as the mother-cell. It is generally deduced from experiments of this nature, that the nucleus is not an essential factor in heredity, even though it may predominate to a fairly wide extent. We must, in fact, bear in mind the arrangement of the cytoplasm, given the anisotropy of the egg; the germs of the characters show a precise layout, just as does the nucleus in the cytoplasm. It is the cell as a whole which determines the transmission of characters, even when for motive power it is inevitable that the nucleus must play a most important part.

PARTHENOGENESIS

In merogony without ovular nucleus we come up against the problem of chromosomic multiplication in the haploid zygote. The same occurs in certain cases of parthenogenesis. On the other hand, natural fertilization results from the *activation* brought about by penetration of the nuclear substance of the male gamete, which immediately dives in and simultaneously sets the division of the female nucleus in motion, the chromosomes of the two nuclei pulling up alongside each other as we have already observed. Nevertheless in certain species and in particular circumstances, division of the egg occurs without the help of the sperm, i.e. parthenogenesis, and furthermore such partheno-genesis can be achieved by artificial stimulation of the nuclear sub-stance of the egg-cell, in what we call artificial or experimental parthenogenesis. These facts have an important bearing from a theoretical standpoint.

Parthenogenesis was demonstrated in aphides by Bonnet in 1845, when ten generations of banana fly were produced without fertilization. Strasburger confirmed the occurrence of parthenogenesis among plants. There are various types of parthenogenesis. Weismann in 1887 thought that the parthenogenetic egg-cell emitted only one polocyte, and that this occurred moreover without meiosis, so that such cell would be diploid. This is not, however, the general rule.

There are several types of parthenogenesis—

(*a*) The so-called *optional parthenogenesis* of certain hymenoptera, xylocopa, bombyx, osmia, apis, hemiptera, the aleurodyde family, acarus, tetranychus, etc., wherein the egg is separated from its two generated polar bodies and remains haploid. From such parthenogenetic eggs we get males exclusively—arrhenotoky—and this generation alternates with sexuate, that is with fertilized offspring, and therefore obviously with diploid zygotes which generate females—thelyotoky. This case is comparatively rare.

(*b*) *Accidental parthenogenesis* is exceptional and of a primitive type, occurring in some echinoderms, asterias and lepidoptera such as the silkworm *bombyx mori*. Only certain of the eggs develop and the remainder degenerate forming two polar bodies with meiosis, though, according to Boveri, Hertwig, and others, it may be possible for the second polocyte to fuse with the egg in a sort of self-fertilization. The egg is diploid.

(*c*) *Cyclic parthenogenesis*, so called because it alternates with sexual reproduction and takes place seasonally only at certain periods in the year. It is the most common type of parthenogenesis. In it, the egg ejects a single polar body and meiosis does not occur; this is the type of parthenogenesis studied by Weismann.

(*d*) *Polyploid parthenogenesis*: the egg commences its chromosomic divisions in the absence of fertilization and division of the cells does not invariably accompany that of the nucleus, so that the resulting cells may well be polyploid. This is to be seen in nematodes and some crustacea. *Artemia* furnishes a typical example: the sexual generation produces specimens with diploid cells of forty-two chromosomes, whilst the parthenogenetic gives polyploid cells with eighty-four chromosomes.

All these classes of parthenogenesis are alike in so far as at a given instant, a female germ-cell starts to divide, thus commencing the embryonic development without participation of a gamete of the opposite sex. Siebold in 1871 compared this process to an internal gemmation, which is furthermore similar to the *apogamy* to be observed in certain plants. The difference depends on the fact that in parthenogenesis the dividing cell, although the reason therefor is not known, is a germ-cell or ovule, whereas in apogamy it is a somatic cell which, through successive division, gives birth to a new individual.

ARTIFICIAL OR EXPERIMENTAL PARTHENOGENESIS

Nothing is known about the mechanism which sets off the multiplication of the ovule. In 1785, Spallanzani made a mixture of lemon juice and vinegar with frog spawn, with the object of stimulating development of the latter. A very important discovery was that of J. Loeb in 1899, namely *experimental parthenogenesis,* or the production of embryos starting with the ovule and without the intervention of a male gamete. The procedure followed originally by Loeb consisted in raising the osmotic pressure of the sea water in which eggs of the arctic urchin *Strongylocentrotus* were immersed. Afterwards, he added fatty acids to the water and placed the ovules straight into the high-tensioned sea water. Delage in 1901 also obtained parthenogenetic development by alternate action of dilute bases and acids, whilst Lefevre in 1907 used dilute organic acids, and other researchers had recourse to physical agents such as heat, X-rays, and the like, or mechanical effects like stirring, shaking and pricking.

Positive results have been obtained under such tests for experimental parthenogenesis in amny species—echinoderms, nemertinea, gephyrea and other annelids, certain fish and batrachia. It has also been possible to activate the egg-cells of certain mammals. Every species is specially sensitive to its particular type of stimulus. Normally, according to Bataillon (1900–27) parthenogenetic development, especially in frogs, comes about after ejection of the polar bodies and the cells come out obviously haploid, which, says Wilson in 1925, serves to explain why many of the larvae obtained by experimental parthenogenesis never succeed in reaching the adult stage. The presence of diploid nuclei in certain cases of blastomeric development does not however appear to be exceptional. Adult forms are sometimes obtained, as Loeb proved even in the frog itself, and it has been stated that frogs can thus be born having a female parent but no male (Curtis and Guthrie, 1928). The biological significance of all these results is very evident.

MATERIAL FACTORS IN DEVELOPMENT

We have just seen that various substances can foster parthenogenetic development of the egg. Equally material agents are those which ensure the attraction of both gametes. Pfeffer in 1897 examined the action of malic acid, whilst Lillie in 1914 showed that the ovule secretes a substance, which he called "fertilisin," the effect of which is to impel the sperm to the ovule and to act as the agent causing penetration by such sperm. Evidently, from this, material factors of a distinct type promote and govern development of the embryo—hormones, harmosones and parahormones—in much the same way as the growth and regeneration of mutilations is developed under the influence of agents of this category. These are the "auxins" or "auxetics" of Rous and Drew, Centanni's "blastins," Fischer's "atraxins," Haberland's "wound hormones," or Caspari's "necrohormones," Fischer's "desmons," Carrel's "trephones," the older "erguse" and "archuse," biotin, the effects of which have been carefully studied. There are, on another level of vital integration, Spemann's "organizers," also Needham's and Waddington's "evocators." The genes too in their turn may well be material factors. Morgan in 1928 questioned whether the gene might reach the size of some of the large organic molecules, e.g. polypeptids, proteases. He asked himself if the organizers might be cyclic hydrocarbons, phenanthrenes, like certain of the cancer-producing substances, sterols, like some morphogenic hormones, carbohydrates, and so on. It seems more probable that the nature of such agents approaches more to a plasmic supramolecular structure than to a collection of isolated molecules.

Nevertheless, chemical substances of very simple molecular composition such as indolacetic acid, or paradichlorophenoxyacetic acid, amongst others, show intense harmosonic properties on the development of plants. A number of very distinct bodies, of which a typical example is represented by certain amino-acids, such as lysin, tryptophan, para-aminobenzoic acid, some hormones and specific vitamins, also influence growth in animals and the development of the organs. Chemical effects in the most diverse classes are responsible for ontogenic development. Indeed starting from the first divisions, this is accurately proven by the study of experimental parthenogenesis. Through the results thereof these general biological principles approach those of physiology and are the commencement of the scientific study of the determinant conditions governing them.

It is necessary to get rid of the preconception that development of the embryo is something specially mysterious or magically different from other organic functions. Growth through assimilation is but one of many functions, and furthermore, the growth and evolution throughout life of an individual, with its characteristic cytological features at any given instant, is in its essence exactly the same thing as embryonic development. This development does not cease when the individual reaches its adult stages; it commences when division of the egg-cell is started and is arrested only by death. Existence is an uninterrupted morphogenic track through time, in the same way that it is also a continuous functional track. Nowadays there is no ground for separating function from form.

* * * * *

EDUARD VAN BENEDEN 1846–1910

Recherches sur la Composition et la Signification de l'Oeuf (Investigations into the Make-up and Nature of the Egg), Brussels: *Mémoires couronnés et mémoires de savants étrangers publiés par l'Académie Royale des Sciences, des Lettres et des Beaux-Arts de Belgique,* Vol. 34, 1868.

It was only after the appearance of the remarkable works of Baer, Purkinje, R. Wagner, Coste, Prévost, Dumas and Rusconi on the vertebrates, and of Rathke, Hérold, Siebold and P. J. van Beneden on the lower animals, that the foundations of comparative embryogeny and ovology were definitely established. The nature of the egg of the higher animals and of a certain number of lower creatures became known, and the notion gradually arose that in all cases the egg was made up of the same essential parts, namely: a membrane, a yolk, and a germinal vesicle containing one or more refringent corpuscles in suspension. Furthermore, the segmentation in batrachia detected by Prévost and Dumas had just been discovered in fish by Rusconi and Baer, whilst Siebold had recently confirmed it in nematoda, and Dumortier, Beneden and Windischmann in gastropoda.

What a mystery this segmentation is, occurring with similar characters in all cases! What relationship could it have with the formation of the embryo, and what was its purpose? This was the seemingly unfathomable puzzle and nobody knew the reason for the yolk segmenting any more than they knew why the egg should contain a vesicle fated to fade away. The real nature of the egg was still beyond our ken.

In 1839, however, Schwann's treatise was published, and the discovery that all animal tissues originate in cells worked a radical change in scientific thought. Schwann's cellular theory threw, perforce, an entirely new light on anatomy and physiology, just as much as it did on embryogeny and was enough to make its author's name famous for evermore.

The cellular theory was now set on a firm foundation, and simultaneously the deep mist hitherto surrounding the question of the nature of the egg and the purpose of segmentation vanished. From his study of its composition, Schwann was the first to proclaim that the egg is a cell; and as all tissues are made up only of cells, it became clear that the purpose of segmentation was to multiply the egg-cell. Bergmann, Reichert and Remak made their particular contributions by showing the part played by segmentation in the production of cells.

Although a great number of physiologists, following Schwann's lead, regarded the egg as a simple cell, others, like Henle, looked on it as an agglomeration of cells, and saw in the germinal vesicle a cell complete in itself. Among those of this opinion we may mention Bischoff, Steinlein, Stein, and others.

It is essential to keep a middle course between these two opinions, and to consider certain eggs as simple cells whilst regarding others as complex. Is it possible, for instance, to consider the threadworm (*trematoda*) and tapeworm (*cestoidea*) eggs as single cells, when they are seen to be formed by the union inside a common shell of a protoplasmic cell with others formed by clearly different glands, erroneously bearing the name of yolk-formers (*vitellogenes*)?

Our own study of the method of formation of the egg and primary embryonic manifestations gives a plain answer to this question.

In every egg, whether it be from mammal or bird, crustacean or trematode, we find a protoplasmic cell whose germinal vesicle is its nucleus, the touch or Wagner's corpuscle, the nucleolus. This cell, which we have termed the germ or egg-cell, and which may be considered as being the original cell of the embryo, *is in all cases formed the same way; it always exhibits the same characteristics and gives birth, by division, to the original cells of the embryo.*

The egg yolk is, however, made up of two elements, one protoplasmic representing the main substance of the egg-cell, and the other, serving as food, forming what we have called the egg's *deutoplasm.*

This deutoplasm is the portion subsidiary to the yolk, and sometimes it is missing, or is produced in a different manner, exhibiting very irregular relation to the protoplasm, and behaving with extreme diversity in the course of the first appearance of the embryo. Often it comes to life right inside the protoplasm and is fashioned by the egg-cell itself; at others, it is formed by special cells either in an exclusively deutoplasm-forming gland, or in the same gland where the germs are formed though in a particular portion of such organ.

This deutoplasm may be formed from specific cells and see the light in a special organ; and though it forms part of the *egg*, it cannot be regarded as an integral part of the *egg-cell*.

It follows that the generally accepted thesis "every egg is a cell" does not bear that stamp of precision which should be a feature of every scientific principle.

But *in every egg there exists one* EGG-CELL, *one germ which is the original cell of the embryo.*

Besides this cell, there resides in the egg a supply of food material, which can be found mixed in with the protoplasm of the egg-cell, and formed in its interior, as may be observed in many vertebrate creatures. In this case they may be included in the cell, and it may be said in agreement with Schwann, who had the vertebrates particularly in view, that *the egg is a cell*.

If, however, the deutoplasm is discovered outside the egg-cell, it cannot be considered as forming an integral part of the germ, and may itself be composed of cells, as can be seen from examples in many of the lower animals which are distinguished by their extreme fecundity. In such case evidently the egg is not one cell but an agglomeration of cells.

THEODOR BOVERI 1862–1915

Das Problem der Befruchtung (The Question of Fertilization), Jena: Fischer, 1901.

In conjugation the chromatin content of two similar nuclei is united. Not only does one of the nuclei consist of the same number of chromosomes as the other, but it can be shown in many cases, where observation has been possible, that the chromosomes from the sperm have the same size, shape and identical properties as those from the ovule nucleus. We may therefore state that each sperm chromosome meets its mate in the female ovular nucleus.

If we denote the chromosomes from one of these nuclei by the letters *a, b, c, d, . . .* , the other nucleus will also contain a similar series *a, b, c, d, ...* These two "haploid" nuclei, on uniting in conjugation, will form a "diploid" nucleus containing *2a, 2b, 2c, 2d, . . .* chromosomes, which we will call the nucleus of the zygote. The double diploid series will develop into identical chromosomes when the zygote divides into two blastomeres and likewise in every subsequent division, since, as is well known, each of the chromosomes forming the matrix splits lengthwise so as to form two daughter matrices in mitosis.

When the nucleus of each cell comes out from the resting stage, the chromosomes become visible as individualized elements always equal in quantity. It is therefore probable that, even during the resting or reproductive period of the nucleus, the chromosomes are at least potentially in existence. The number of chromosomes is characteristic of the particular animal or plant species. Furthermore, as we have just seen, in mitotic division the chromosomes are equally divided and become paired off during fertilization, so that we may feel justified in contending that the chromosomes possess "individuality."

In blastomeric division after fertilization the two series, male and female, are transmitted in exact balance, making up mated pairs of elements from each source. This process is repeated indefinitely throughout all the cell divisions which take place during development and subsequently throughout the whole course of existence.

This is made possible precisely because the chromatic reduction, taking place during the ripening period of the sex cells by reason of the haploid character acquired by these cells, brings in its train the natural result that the fertilized ovule or zygote is in its turn diploid. Such homogeneous distribution of chromosomes to all the cells in the body is only feasible on the condition that the mitotic forms are normal, regular and bipolar. Mitosis with three or more irregular poles is accompanied, as may well be understood, by a heterogeneous arrangement of the chromosomes, unsuitable for normal development. (*See* also: paper "On the development of malignant growths," *Zur Frage der Entstehung maligner Tumoren,* Jena, 1914.)

EDMUND BEECHER WILSON 1856-1939

The Cell in Development and Heredity, New York: Macmillan, 1924.

In fixed material, especially as viewed under relatively low magnification, the nuclei commonly appear deeply stained after treatment by certain dyes, such as carmine, haematoxylin, methyl-green, or gentian violet, while the cytoplasm remains relatively pale. Such dyes, accordingly, are often designated as "nuclear dyes," in contradistinction to the "plasma-dyes" which stain especially the cytoplasmic substance. The cytoplasm often contains various formed elements (granules, fibrillae, etc.) that may likewise be deeply stained by the "nuclear" dyes. The term "plasma-dyes," therefore, only denotes their predominant effect on the cytoplasm considered as a whole; examples of the latter are offered by eosin, acid fuchsin, orange G or light green. It was shown by Ehrlich (1870–80) and his successors that the nuclear dyes in general and, in particular, the aniline dyes or coal-tar colours, are "basic," the plasma-dyes "acidic," and it is convenient, accordingly, to designate the various cell-components as *basophilic* and *oxyphilic* according to their tendency to take up the basic or the acidic dyes. On what this tendency depends—whether on chemical affinity, on physical processes of adsorption, or on both —need not here be considered.

The earlier cytologists, employing for the study of the nucleus mainly the basic or nuclear dyes (especially carmine, haematoxylin, and later safranin and gentian violet) observed that in fixed material, and after certain technical manipulation, only certain components of the nucleus were stained by these dyes. To the substance thus stained Flemming (1880) gave the name of *chromatin*, to that which stains slightly or retains the colour feebly upon extraction (by acids, etc.) *achromatin*. "Chromatin," as thus defined, was considered by Flemming to be composed wholly or in part of the chemical substance "nuclein" and to form the more conspicuous part of the nuclear framework and also certain types of nucleoli. Under the conception of "achromatin" Flemming included all the remaining nuclear substance except the enchylema. Strasburger (1882) and Carnoy (1884) recognized that the framework itself appears to consist of two constituents, namely, a continuous "achromatin" basis, and of more or less discontinuous granules or clumps of "chromatin" suspended in it. The first of these was found to be oxyphilic and was accordingly designated by

Strasburger as *nucleo-hyaloplasm*, by Carnoy as *plasmatic network* (composed of "plastin") and later by Schwartz (1887) as *linin*, a term still in common use. To the foregoing differences may be added the fact that "chromatin," as thus defined, shows a high degree of resistance to hydrochloric digestion, while the oxyphilic "linin" is less resistant in varying degree. . . .

The nuclear division is of two widely different types, which came to be known as *direct* and *indirect* (Flemming, 1879). In the direct and simpler type the nucleus, like the cell-body, undergoes a simple mass-division into two parts. In the indirect and more complex type, the nucleus is not destroyed, but is spun out into long threads which *split lengthwise* so that every portion is exactly divided between the daughter-nuclei. Before the separation of their longitudinal halves these threads shorten, and thicken to form more condensed bodies known as *chromosomes* (so named by Waldeyer because of their intense staining-capacity). The products of their fission (daughter-chromosomes) separate, and pass to opposite poles; and from the two groups of daughter-chromosomes are rebuilt two corresponding daughter-nuclei which by reason of the preceding processes are exact duplicates of each other and of the mother-nucleus. In this operation we now recognize one of the most fundamental mechanisms of heredity.

By Schleicher (1878) this process was called *karyokinesis*, a term still widely employed for cell-division of this type. Flemming (1882) proposed the more appropriate term *mitosis*, in allusion to the characteristic thread-formation, while the direct mode of division was called *amitosis*; and this usage gradually became firmly established. Other terms are *karyodieresis*, *cytodieresis* (Henneguy), *kinesis* and *akinesis* (Fol, Carnoy), but these are less generally used. Strictly speaking, all these terms refer to division of the nucleus, but by an extension of meaning they are often applied to cell-division as a whole. It is often convenient to employ the term *cytokinesis* (Whitman, 1887) to designate the associated changes taking place in the cytoplasmic cell-body, though in practice it is sometimes difficult to draw any definite line of distinction between the nuclear and the cytoplasmic activities, e.g. in the formation of the spindle. Cytokinesis includes not only the division of the cytosome as a whole but also the orderly distribution of smaller elements within it; such as the chondriosomes of the Golgi-bodies; and these processes have received corresponding names (*chondriokinesis*, *dictyokinesis*).

EDMUND BEECHER WILSON 1856–1939

The Cell in Development and Inheritance, New York: Macmillan, 1896.

The splitting of the chromosomes is, in Boveri's words "an independent vital manifestation, an act of reproduction on the part of the chromosomes."

All of the recent researches in this field point to the conclusion that this act of division must be referred to the fission of the chromatin-granules or chromomeres of which the chromatin-thread is built. These granules were first clearly described by Balbiani (1876) in the chromatin-network of epithelial cells in the insect ovary, and he found that the spireme-thread arose by the linear arrangement of these granules in a single row like a chain of bacteria. Six years later Pfitzner (1882) added the interesting discovery, that during the mitosis of various tissue-cells of the salamander, the granules of the spireme-thread *divide by fission and thus determine the longitudinal splitting of the entire chromosome*. This discovery was confirmed by Flemming (1882), and a similar result has been reached by many other observers. The division of the chromatin-granules may take place at a very early period. Flemming observed as long ago as 1881 that the chromatin-thread might split in the spireme stage (epithelial cells of the salamander), and this has since been shown to occur in many other cases; for instance, by Guignard in the mother-cells of the pollen in the lily (1891). Brauer's recent work on the spermatogenesis of *Ascaris* shows that the fission of the chromatin-granules here takes place even before the spireme stage, when the chromatin is still in the form of a reticulum, and long before the division of the centrosome. He therefore concludes: "With Boveri I regard the splitting as an independent reproductive act of the chromatin. The reconstruction of the nucleus, and in particular the breaking up of the chromosomes after division into small granules and their uniform distribution through the nuclear cavity, is, in the first place, for the purpose of allowing uniform growth to take place; and in the second place, after the granules have grown to their normal size, *to admit of their precisely equal quantitative and qualitative division*. I hold that all the succeeding phenomena, such as the grouping of the granules in threads, their union to form larger granules, the division of the thread into segments and finally into chromosomes, are of secondary importance; all these are only for the purpose of bringing about in the simplest and most certain manner, the transmission of the daughter-granules to the daughter-cells." "In my opinion the chromosomes are not independent individuals, but only groups of numberless minute chromatin-granules, which alone have the value of individuals."

These observations certainly lend strong support to the view that the chromatin is to be regarded as a morphological aggregate—as a congeries or colony of self-propagating elementary organisms capable of assimilation, growth and division. They prove, moreover, that mitosis involves two distinct though closely related factors, one of which is the fission of the chromatic nuclear substance, while the other is the distribution of that substance to the daughter-cells. In the first of these it is the chromatin that takes the active part; in the second it would seem that the main role is played by the archoplasm, or in the last analysis, the centrosome.

Reproduced by permission of the publisher

THOMAS HUNT MORGAN (1866–1945), ALFRED HENRY STURTEVANT (*b.* 1891), HERMANN JOSEPH MULLER (*b.* 1890) *and* CALVIN BLACKMAN BRIDGES (1889–1938)

The Mechanism of Mendelian Heredity, New York: Holt, 1915.

THE CHROMOSOMES AS BEARERS OF HEREDITARY MATERIAL

The evidence in favour of the view that the chromosomes are the bearers of hereditary factors comes from several sources and has continually grown stronger, while a number of alleged facts, that seemed opposed to this evidence, have either been disproven, or else their value has been seriously questioned. . . .

It has been argued that since the sperm transmits equally with the egg, and since only the sperm head, consisting of the nucleus, enters the egg, inheritance is only through the nucleus. But it must be admitted that around the entering sperm nucleus there may be a thin enveloping protoplasm, which, however scanty, might suffice to transmit certain cytoplasm factors. Moreover, while the tail of the sperm appears in some cases to be left outside the egg, in other cases it appears to enter and to be absorbed.

Behind the head of the spermatozoön, and at the base of the tail, there is a middle piece which contains a derivative of the old centriole or division centre. Since the centrosome carried by the sperm has been found in some forms to give rise to the new centrosomes that occupy the poles of the first cleavage spindle of the egg, it may appear that a paternal contribution can come about in this way. It is true that the continuity of the centrosome of the sperm with that of the dividing egg has been disputed in some forms; but it is difficult to prove that the sperm centrosome is lost, even though it may disappear owing to loss of staining power.

The nucleus contains a sap which is probably of cytoplasmic origin. The presence of this sap may again be appealed to by those who do not accept the chromosomes as the bearers of heredity, as a weak link in the evidence. It is true that the nuclear sap appears to be squeezed out of the nucleus of the sperm head, leaving a compact and apparently solid mass of chromatin, yet its complete elimination can not be proved. Hence, while those who favour chromosomal transmission find in the facts of normal fertilization strong *indications* favourable to that view, yet it is also true that those who are inclined to dispute this view find several loopholes in the arguments of their opponents.

The importance of the nucleus in heredity has further been shown by experiments of Bierens de Haans, Herbst and Boveri on giant eggs of sea-urchins fertilized by sperm of another species. The hybrid larvae produced when normal eggs of one species are fertilized by sperm of the other species are intermediate in character between the two parental types of larvae; while those from giant eggs of the same species fertilized by sperm of the other, also intermediate, incline more to the maternal side. The nucleus of the giant egg is double the size of that of the normal egg, and according to Bierens de Haans, the chromosomes are also double in number. Consequently the amount of maternal chromatin should be double that introduced by the sperm, and might produce a corresponding influence on the hybrid character. But since in these giant eggs the cytoplasm is also doubled it is not evident that the results are due to the chromosomes rather than to the cytoplasm.

Boveri's studies upon dispermic fertilization of the egg of the sea-urchin bear directly upon the question at issue. He found that when two sperms simultaneously enter the same egg, each brings in a centrosome so that a tetra- or tri-polar spindle is formed for the first division. Instead of a double set of chromosomes, as in normal fertilization, there are three sets. At the first division, the chromosomes are irregularly distributed upon the multipolar spindles. In consequence, some cells may get one of each kind of chromosome, while other cells may get less than a full complement. These dispermic eggs almost always give rise to abnormal embryos, as several observers have recorded. The result can best be attributed to the irregular distribution of qualitatively different chromosomes; only those embryos in which each cell has a full complement developing normally.

The view that the chromosomes are persistent as individual structures in the cell has steadily gained ground during the last twenty years. The process of karyokinetic division by means of which at each cell division the halves derived from a lengthwise split of each chromosome are carried to opposite poles, so that a genetic continuity is maintained between corresponding chromosomes (and parts of chromosomes) in mother- and daughter-cells, has been found to be almost universal in both plants and animals.

During the resting state the chromosomes spin out in such a way that they appear to form a continuous network in the nucleus. They cannot be identified individually during this period. When the chromosomes again become visible, preparatory to the next division, it has been found by Boveri in *Ascaris*, which is particularly well suited for the study of this point, that in sister-cells the configuration of the groups of chromosomes is the same. The similarity of the sister-cells would be expected had the chromosomes retained during the resting stage the same shape and size and relative location that they had at the end of the last division.

In many animals and in some plants the chromosomes are of very different sizes and shapes, and many, or even all of them, can be identified at each division. It is found that these size relations hold throughout all divisions of the cells. While this evidence appears at first sight to show that the chromosomes are structures that perpetuate themselves, preserving their identity, yet it might be maintained, in fact it has been maintained, that each species has its own peculiar protoplasm from which chromosomes of a particular kind and number are, as it were, crystallized out anew before each cell division. This point of view cannot however be reconciled with the evidence that follows.

In general, it may be said that even an abnormal set of chromosomes, once established in a cell, tends to persist through all succeeding cell generations. This evidence indicates that the chromosomes are not mere products of the rest of the cell but are self-perpetuating structures.

CHAPTER IX

HEREDITY

HEREDITARY TRANSMISSION

CHILDREN resemble their parents, and the characters of the species are transmitted from generation to generation. Darwin in 1859 stated that heredity must be subject to natural laws at that time as yet unknown. He noted that if individuals of different races be crossed, their descendants will exhibit combined and intermediate characters which in the course of succeeding generations tend to revert to their original forms, that is, revert to type.

It is common knowledge indeed that cross-breeding gives rise to offspring exhibiting characters from the male line in combination with others from the female. The distribution of these characters from different sources is not constant in all cases, though the offspring will always exhibit morphological and functional elements, which have descended to it from its two parent strains. Sometimes the influence of one parent will predominate greatly over the other, giving what we call *alternating heredity*, whilst in others and indeed the greater number of cases the form of the child is *intermediate* or halfway between that of the two parents, either in *mosaic* where the interpolation of the paternal and maternal characters is visible, or in *blending* or *fusion*. Naudin in 1863 crossed individuals of different races and even of different species and showed that offspring of the first generation were usually intermediate between their two progenitors, bearing greater resemblance to the male parent in some cases and to the female in others, though in general there seemed to be a balanced distribution of characters. Naudin also pointed out a reversion to original type of one of the parental lines releasing the offspring from the influence of the least dominant progenitor after a series of generations.

QUANTITATIVE LAWS OF HEREDITY

Francis Galton, who in fact was a cousin of Darwin, studied the phenomena of heredity from the statistical angle in accordance with methods which had been proposed by Quetelet in 1845. In 1889, Galton was in a position to formulate the *quantitative laws* of heredity, according to which the parents contribute in equal shares a half of the characters exhibited by the child; the four grandparents contribute jointly a quarter, i.e. a sixteenth part each; the eight greatgrandparents are responsible for an eighth altogether, and so on and on. Pearson, Davenport and others confirmed and elaborated these findings, proposing various algebraic formulae to express them.

Such participation by each individual in the ancestral chain, which, as we see it, keeps on diluting with the succession of generations, explains the tendency to revert to specific type, as expressed by Galton's second law. The offspring inherits more from its immediate ancestors than from the more remote, but heredity tends to erase individual differences and return any individual to the common mass of the population or to a merely statistical mediocrity. This in itself implies the existence of the species which, precisely because it is a species, is evidence of a certain constancy.

The science of the laws of heredity underwent great progress thanks to the labours of Gregor Johann Mendel. In his paper on cross-breeding experiments *Versuche über Pflanzenhybriden* in 1865, which at the time went practically unnoticed, Mendel gives a summary of his researches, and his main conclusions were—

(*a*) Characters transmitted by heredity do not mingle, but maintain their particular quality without blending with the characters imparted by the other parent; this is disjunction of characters, i.e. characters from the male parent and from the female parent are not added to one another or blended into a compound.

(*b*) As the paternal and maternal influences coexist and one only of them is evident in the pairs of characters or allelomorphs from each of the lines, male and female, there must be a *dominant* character giving visible effects and a *recessive*. The characters are arranged in pairs, such as the chromosomes, or the genes, but from the pair for each one of the characters, only one element appears active; the other lies dormant and will be effective only when there is no longer a dominant coexisting with it to overshadow it and prevent it from making itself evident.

Mendel effected crosses between races of the same species, thus obtaining infinitely fertile series of specimens which allowed him to study a great number of generations. He called these crossings *hybridisms*, and not as would have been correct *cross-breeds*, and the offspring obtained by the crossing of two different progenitors were called hybrids, whether from the same species or from different species.

Mendel got easily evaluable results from his researches, since he chose a few well-defined, readily observable, differentiable characters. His first observations were made with pea vines *Pisum sativum*. For instance, he crossed tall standards with short ones, and, in other experiments, plants giving smooth round peas with those giving wrinkled peas, green peas with yellow peas, sweet with tasteless, plump pods with shrivelled ones, and so on. Having selected the two kinds he wished to cross, he prevented their self-fertilization in the flower, and on the contrary fertilized the blossoms of the one variety with pollen from the other, denoting the generation by the letter P (parents). Afterwards, he sowed the seeds obtained and observed the characters of the *first generation* of descendants (F^1). He then allowed the individuals of this generation to fertilize themselves and sowed the seeds thus obtained, thereby getting the *second generation* (F^2), and so on.

Crossing an individual from the tall variety with one from the short, all the F^1 individuals were tall without any noticeable difference. Thus the character "tall" must be dominant and "short" recessive. In the second generation of the progeny, which are hybrids, F^2, we get groups of four offspring, three tall and one short, or 75 per cent dominant individuals and 25 per cent recessives. Of course, working with a small number of specimens, the ratio will be only approximate, but mathematically and statistically it will become more and more exact as the number of individuals increases in total.

Self-fertilization of recessive standards constantly gave short, recessive offspring. On the other hand, self-fertilization of the tall plants gave different results—one-third of them produced dominants exclusively which on self-fertilization again conserved their dominant characteristic without exception; the other two-thirds were hybrids repeating the F^1 experience, namely groups of four with two pures, one dominant and one recessive, and two hybrids.

This quantitative relationship was repeated on trial with other characters and also using other species of plants, such as kidney beans, wheat, Indian corn, barley, oats, and so on, as well as with animals, e.g. mice, rabbits, guinea-pigs, etc., so that it must therefore be a general law.

Since self-fertilization of hybrids gives birth to other hybrids, though these include also a *pure* paternal and a *pure* maternal form for every two hybrids, it may be deduced that the factors determining the characters are not fused but are *disjoint* or *separate* in the germ-cells or gametes. Accepting, for the purposes of simplifying the argument, that a determinate factor, a gene perhaps, defines a certain character and that the genes are arranged in pairs, one gene in each pair being respectively from the male and one from the female parent, a very simple combination of genetic elements in fertilization will give one zygote in which will be found solely elements from one of the progenitors, and another in which occur those from the other parent making up the *homozygote* offspring. In the other two specimens we shall find the paternal element jointly with the maternal giving *heterozygote* offspring or half-castes exhibiting characters imposed on them by the dominant element.

	A	*a*	ovules
A	*AA*	*Aa*	
a	*aA*	*aa*	

pollen

The zygotes will be 1 *AA*, 2 *Aa*, 1 *aa*. But since *A* (tall) is dominant over *a* (short), the offspring will be three talls and one short. The three talls are not, however, identical, they correspond to the same *phenotype* but differ in their germinal constitution, a third of them—one specimen in the case of a single group—being homozygote or pure *AA*, and two-thirds being made up of hybrids *Aa*, i.e. heterozygotic, giving tall plants since that character is dominant and overclouds the recessive character, though the gene of such recessive is also present. From this it is obvious that the tall phenotype will be made up of two different *genotypes*, the pure and the hybrid, which can be distinguished only through the offspring to which they give birth.

The numerical proportions are maintained when, instead of taking into account a single set of characters or alleles, say tall-short, smooth-wrinkled, green-yellow, and so on, we study what happens on crossing varieties with two sets of alleles, for instance yellow, smooth peas with green, wrinkled ones, obtaining thus two dihybrids. Trihybrids can also be obtained by crossing tall specimens having smooth, yellow seeds with short ones having rough, green seeds. And in this manner we can go on progressively complicating the combinations.

In the dihybrids the genetic elements make their appearance

independently as usual, disjunct and separate as shown by the combinations YS and ys, the dominants yellow and smooth, and the recessives green and wrinkled.

Sixteen combinations of gametes are possible through fertilization, since there are four possible distinct classes of sperm as well as four of ovules. Thus the generation F^2 will give four phenotypes YS, Ys, yS and ys in the proportion 9 YS, 3 Ys, 3 yS and 1 ys. This 9 : 3 : 3 : 1 proportion is the same as the 3 : 1 ratio found in the monohybrids and multiplied by 3 in the first two places.

The same occurs with trihybrids. In the F^3 generation, there will be eight phenotypes and the proportion of the number of individuals will be as 27 : 9 : 9 : 9 : 3 : 3 : 3 : 1. The same mathematical laws hold, however intricate the combinations become.

STATISTICS AND CYTOLOGY

The importance of the facts laid bare by Mendel was realized only many years afterwards. At first, they seemed to be merely a collection of simple numbers, but later on cytological science was seen to show that these figures tallied with the make-up of the components of nuclear substance. For some time consequent to 1865, Mendel's investigations remained little less than unknown, but at the beginning of the twentieth century, de Vries in Amsterdam, Correns at Tübingen, and Tschermack in Vienna, all simultaneously arrived at the same conclusions as Mendel when investigating the laws of hybridization. From that time forward, researches connected with the Mendelian laws of heredity have progressed by leaps and bounds, especially in the United States, starting with Bateson in 1903 and proceeding particularly under the leadership of T. H. Morgan from 1910 onwards. Sutton (1903) brought to light the relation between transmission of Mendelian characters and the data furnished by cytology, showing how the quantities observed tally with the behaviour of the chromosomes with their genes, constituting allelomorphs in the cell, which become separated in meiosis by the lengthwise splitting of such chromosomes.

ABSOLUTE AND RELATIVE DOMINANTS

Countless observations which have been made in the present century up to now, with regard to Mendelian transmission, have fully confirmed the two basic laws: (*a*) disjunction, and (*b*) dominance and recession.

It is, however, becoming clear that the influence of the dominant is not complete in every case and therefore, in certain specimens, the hybrid forms a phenotype differing from the dominant, since the corresponding character is added to that of the recessive. Such is the famous case of *Mirabilis jalapa* (the "Four o'clock") among others, where on crossing red-flowered specimens of this species with white-flowered ones, the F^1 hybrids give pink petals, that is to say, petals of a colour halfway between those of the two parents. Thus in F^2 we get one individual with red flowers, another with white and two hetero-zygotic hybrids with pink. Such blending of characters in the hybrids does not signify lack of disjunction, since self-fertilization of the F^2 hybrids results in pure male and pure female homozygotic offspring in accordance with the standard law. This case, which is by no means unique, only goes to show that sometimes the dominant element is not sufficiently in the ascendant as to overcome entirely the influence of the recessive, which state of affairs causes a certain interference in the powers of the dominant and must be taken into account when seeking to interpret certain cases of hereditary transmission.

There arises also the possibility that the recessive character, not remaining latent, may unite with the dominant without forming a more or less uniform blend, but giving rise to blended intermediate characters wherein both dominant and recessive characters alternate in effect to give a mosaic. This is to be observed in shorthorns, where the crossing of white stock with red gives roan offspring. The appearance of this character concurs with Mendelian statistical laws, the roans being heterozygotes and their intercrossing giving offspring in the ratio 1 red, 2 roan, 1 white, which is clear evidence of disjunction of the characters. Another example is that of the so-called Andalusian Blue poultry, to obtain which it is necessary to cross blue-speckled whites with blacks. The F^1 offspring is the blue hybrid and crossing two specimens of this generation gives 1 speckled white, 2 blues and 1 black.

TRANSMISSION OF COMPLEX CHARACTERS

The action of multiple genes may mount up in such a manner that it is difficult to retrace the characters of the original homozygotes, which can be explained by reference to the formulae of the calculus of probabilities applicable to the case. This is particularly noticeable in crossings of the human species. The crossing of white with negro gives a more or less dark mulatto, according to the predominance of the white or the negro characters in one progenitor or the other. The offspring of the two mulattoes will not, however, revert to the pure white or pure negro. In other half-breeds the possible effect of "lethal" germs must be taken into consideration, which by unfailingly bringing about the death of some of the offspring apparently modifies the normal quantitative relationships.

Such examples, among others which we could adduce, show that apparent exceptions to Mendelian statistical forecasts in no wise invalidate the general rule, even though it may be necessary to bear in mind the concatenation of special conditions in such exceptional cases. These problems are, however, practically always of such complexity that their analysis is frequently most difficult. Our knowledge of heredity is today rather more extensive than the ideas Mendel had available to take as a basis for his famous experiments, and the tremendous wealth of material now at our disposal only serves still more to corroborate the accuracy of his original deductions.

LINKAGE AND CROSS-OVER

The dependence of a character with regard to the corresponding genes contained in the chromosome—the intrinsic material basis upon which Mendelian statistics subsist—is shown by the fact of character *linkage*. The characters are independent and referable to particular genes, preserving always their biological individuality. They do not, however, appear or act separately. Some characters carry others in their train.

Goethe in 1780, on discussing the *Metamorphoses of Plants*, even that long ago, mentioned the correlation or subordinacy of various characters. Bernard and Jussieu in 1798 likewise demonstrated the "subordinacy" of characters, namely, that a main character implies the appearance of certain secondary characters. Bateson and Punnet in 1905 rediscovered such subordinacy, i.e. the mutual dependence of certain characters, which nowadays is termed *linkage*, arising from the presence and more or less close vicinity of corresponding genes in one and the same chromosome. An interesting example of linkage is that of the relations set up between the chromosomes which determine sex and certain cognate characteristics.

The same system of charting which helps to explain linkage allows us also in many cases to interpret the *cross-over* of certain characters. "The presence of associated genes," says Woodruff in 1946, "in the same chromosome by no means indicates that these genes must *always* be distributed together. Thus during synapsis, homologous genes (alleles) often reciprocally CROSS OVER from one synaptic mate to the other, and so become separated from their former gene associates in the same chromosome." Consequently, when chromosomes separate through division of the equatorial plate or mother aster, the descendant chromosomes may show alternation of genes and the characters of the offspring will modify the results accordingly. The cross between two homologous chromosomes may be unique or variable and in this way *recombinations* of genes may come about, to which will correspond recombinations of characters. This obviously grants a certain flexibility to the genetic mechanisms. The study of so important a phenomenon has again confirmed the justice of attributing characters to their chromosomic elements, and has consolidated the basis of the Mendelian theory, imparting to it the necessary flexibility, and facilitating the elaboration of admirable gene-maps which show accurately in each chromosome the spot where, theoretically, the various genes must lie. Theory is moreover justified by practical results. In few other domains of science has it been possible to find so perfect a correlation between theory and actual practical results. Facts demonstrated in cytology help us to interpret the Mendelian laws, though all this does not mean to say that the gene hypothesis alone can explain the development of living organisms, any more than the assumption that a phenomenon may be due to the effect of a particular factor, or of several factors more or less hypothetical, will explain such particular phenomenon in any of the realms of biology.

With these reservations, we can however assert that knowledge of the conditions leading to transmission of hereditary characters has made extraordinary progess, and so great has been the advance that in many cases it is possible to foretell the forms that offspring from certain given progenitors will take, as well as the arithmetical proportions of different types of such progeny. Genetics is rapidly becoming an exact science and is profiting from the facts we have been able to analyse statistically, even when we do not know—as indeed no science really can know—exactly what those facts are in their intrinsic significance.

* * * * *

GREGOR JOHANN MENDEL 1822–84

Versuche über Pflanzenhybriden (Plant Breeding Experiments), *Verhandlungen des Naturforschenden Vereins in Brünn*, Vol. 4, 1865.

Numerous experiments have shown that if two plants differing constantly in one or more characters be crossed with each other, the common characters will be transmitted without change to their hybrid progeny, but that each pair of different characters will on the other hand combine together to form a new and usually variable character in the offspring. The object of my experiments was to observe such variations in connection with pairs of different characters and to deduce the laws governing their appearance in succeeding generations.

Various forms of peas were selected for crossing, showing differences in—

   (i) Length and colour of stem.

  (ii) Size and shape of leaves.

 (iii) Position, colour and size of flower.

 (iv) Length of flower stalk.

  (v) Colour, shape and size of pod.

 (vi) Shape and size of seed.

(vii) Colour of seed-coat and cotyledons.

There are some characters which do not allow of a sharp and clear distinction, since their differences are merely a matter of degree often difficult to define, and they cannot therefore be used in these particular researches which of their very nature had to be confined to features allowing of no confusion.

The particular characters selected for experiment were accordingly—

1. *Differences in Shape of the Ripe Seed*

These are either round or roundish with shallow wrinkles on the surface, or irregularly angular with deep wrinkles.

2. *Differences in Endosperm Colour*

The ripe seed endosperm ranges from pale yellow or bright yellow to orange, or else possesses a more or less intense green tint. This variation is easily visible in the seeds since their coats are transparent.

3. *Differences in Seed-coat Colour*

This is either white, when it is invariably in correlation with white flowers, or grey, grey-brown or leather-brown with or without mauve spots, in which latter case the colour of the standards will be mauve, that of the wings purple, and the stem of the leaf-axils reddish. The grey seed-coats turn dark brown in boiling water.

4. *Differences in Shape of Ripe Pods*

These are either smooth and plump or more or less wrinkled and deeply constricted between the seeds.

5. *Differences in Colour of Unripe Pods*

These are either in the range from light to dark green or else of a bright yellow, the stalks, leaf-veins and calyx partaking of the same colour.

6. *Differences in Position of the Flowers*

The flowers are distributed either axially along the main stem or bunched together at the top, in which latter case the upper end of the stem is more or less wide in section.

7. *Differences in Length of Stem*

The length of stem varies considerably in certain plants, but is nevertheless a constant character for any particular species in so far that healthy plants grown in the same soil are subject to merely unimportant variations in this character. In experiments dealing with this character, long stems of 6 to 7 ft were crossed always with short ones of 9 in. to 1 ft 6 in. so as to allow certainty in discrimination.

Each pair of the different characters listed above was united by cross-fertilization, in accordance with the following table—

1st	test,	60	fertilizations	on	15	plants
2nd	„	58	„	„	10	„
3rd	„	35	„	„	10	„
4th	„	40	„	„	10	„
5th	„	23	„	„	5	„
6th	„	34	„	„	10	„
7th	„	37	„	„	10	„

From a quantity of plants of the same variety only the most vigorous were selected for fertilization. Ailing plants will give only uncertain results since, even in the first generation of hybrids, and still more so in subsequent ones, many of the offspring will either fail entirely to flower or will form only a few seeds of inferior quality.

FORMS OF HYBRIDS (F^1)

Hybrids are not, as a rule, exactly midway between their parental species. This certainly holds for pea hybrids. In each of the seven crosses indicated above, the hybrid-character is so near to that of one of the parental forms that the other escapes observation entirely or at any rate cannot be detected with certainty. Those characters which are transmitted entire or with hardly any change on fertilization, and are consequently in themselves the main characters of the hybrid are termed *dominant*, while those which become latent are called *recessive*; this latter expression has been chosen because the characters thus designated withdraw or disappear to all intents and purposes in the hybrids, even though they make their appearance again unchanged in the progeny of such hybrids, as will be shown further on.

These experiments showed, furthermore, that it is absolutely immaterial whether a dominant character belongs to the seed-bearer or to the pollen-parent; the form of the hybrid remains identical in either case. This interesting fact was also confirmed by Gärtner, who stated that even the most practised expert cannot determine in a hybrid which of its two parent species was male or which female.

Among the differentiating characters employed in these experiments, the following are dominant—

1. Round or roundish seed with or without shallow wrinkles.

2. Yellow colour of endosperm.

3. Grey, grey-brown or leather-brown colour of seed-coat, in conjunction with purple blossoms and reddish spots in the leaf-axils.

4. Smooth, plump pod.

5. Green colour of pod, with same colour in stems, leaf-veins and calyx.

6. Flowers distributed along the length of the stem.

7. Long stem.

THE FIRST GENERATION BRED FROM THE HYBRIDS (F^2)

Recessive characters with all their peculiarities reappear in this generation alongside the dominant characters, in the definite average proportion of 1 : 3, so that in every four plants of this generation, three show the dominant character and one the recessive. This applies without exception to all the characters covered by the experiments; in no case were any transitional forms observed.

The relative quantities of each pair of differentiating characters obtained were as follows—

Test 1: Shape of Seed

During the test year under consideration, 7,324 seeds were obtained from 253 hybrids, giving 5,474 round or roundish seeds and 1,850 angular, wrinkled ones, whence we get the ratio 2·96 : 1.

Test 2: Endosperm Colour

258 plants yielded 8,023 seeds of which 6,022 yellow and 2,001 green, giving the ratio of 3·01 : 1.

Test 3: Seed-coat Colour

Of 929 plants, 705 bore purple flowers and grey-brown seed-coats, whilst 224 had white flowers and white seed-coats, giving the ratio 3·15 : 1.

Test 4: Pod Shape

Out of 1,181 plants, 882 were plump and 299 constricted, whence the ratio 2·95 : 1.

Test 5: Pod Colour

580 plants were submitted to test, of which 428 had green pods and 152 yellow ones, or a ratio of 2·82 : 1.

Test 6: Position of Blossoms

Among 858 specimens, 651 blossoms were axial and 207 terminal, giving the ratio 3·14 : 1.

Test 7: Length of Stem

Out of 1,064 plants, the stem was long in 787 cases and short in 277, whence the ratio of 2·84 : 1.

The dominant character may here possess a double significance representing a pure parental character or a hybrid character. In which particular aspect it appears in any particular case can only be determined in the succeeding generation. As a parental character it will pass over unmodified to all offspring, whereas if on the other hand it belongs to the hybrid, it will follow the same behaviour as in the first generation.

The Second Generation bred from the Hybrids (F^3)

Those forms which maintain the recessive character in the first generation do not vary in regard to it in the second generation but remain constant in their offspring. Those which possess the dominant character in the first generation from the hybrids behave otherwise, for from them two-thirds yield offspring displaying dominant and recessive characters in the proportion of three to one, and showing thereby the same ratio as the hybrid forms, whilst *one-third* only behaves as a pure dominant transmitting the dominant character as a constant without modification.

Some of the experimental results are given below—

Experiment 1

Among 565 plants raised from round seeds of the first generation, 193 gave round seeds only, remaining accordingly constant as far as this character is concerned, whereas 372 gave both round and angular seeds in the proportion of 3 : 1. The number of hybrids therefore as compared with the pure constants is 1·93 : 1.

Experiment 2

From 519 plants raised from seeds with yellow endoderm in the first generation, 166 yielded yellow exclusively, whilst 353 gave yellow and green seeds in the proportion of 3 : 1. The division into hybrid and pure forms comes out therefore at 2·13 : 1.

In every experiment a certain number of plants always came out with the dominant character. The above two experiments are of special value in determining the proportion in which the forms with the constantly persistent character occurs. The average of the ratios 1·93 : 1 and 2·13 : 1 taken together gives practically the exact ratio 2 : 1.

The three-to-one proportion governing the distribution between dominant and recessive characters in the first generation, resolves itself therefore over the whole series of experiments into the ratio of 2 : 1 : 1 if the dominant character be differentiated in accordance with whether it be a hybrid character or a pure parental one. Since members of the first generation spring directly from seeds of the hybrids, it now becomes clear that the *hybrids produce seeds containing one or other of the two differentiating characters, of which one-half again give fresh hybrids, whilst half yield offspring which remain constant in their genetic properties and receive the dominant or recessive in equal quantities.*

YVES DELAGE 1854–1920 *and* MARIE GOLDSMITH

Les Théories de l'Evolution (Evolution Theories), Paris: Flammarion, 1916.

The fundamental question in any theory of heredity is that referring to the *physiological process* whereby an organism in course of development acquires similarity to its forebears. The problem may, however, be posed another way, namely, by leaving aside the innermost phenomena occurring within the fertilized egg and the various tissues which make their appearance and become differentiated, at the same time taking for granted the resemblance between parents and offspring, we may consider merely similarities of features, gradations therein and their variations over the course of time. It was along these lines that Galton and Mendel carried out their researches and formulated the laws which we will now proceed to expound.

In the first place, we should mention analysis of hereditary effects in the light of *statistical* methods, for the purpose of deducing certain general laws from observation of a large number of facts. The idea of applying statistics to biological questions and particularly to variational phenomena is due to Francis Galton, who laid the foundations of a new science "Biometry" by his two famous books *Hereditary Genius* (1869) and *Natural Inheritance* (1889). He was soon followed by many scientists such as his direct successor, Karl Pearson, and others like Weldon, Bateson, Darbishire, whose labours were for the most part published in the journal *Biometrika* devoted especially to this subject.

The first general principle Galton was able to establish, on the basis of abundant statistics covering variations relating to 150 families and embracing a huge number of distinguishing physical and mental factors, was that if variations of a particular character or ability be taken into account, it seems certain that a constant average exists for each generation and that differences as compared with such average mutually balance one another. If, for instance, a parent be much taller or shorter than the average, his children have a leaning in the opposite direction. There is a tendency to revert to the average type, since such norm represents the most stable equilibrium.

Heritage does not come solely from the immediate parents, but from all previous ancestors, and their contribution is the greater the closer they are to the particular generation under observation. Galton combined all these considerations into one scheme which he called the "Law of Ancestral Inheritance." Each generation dating back from the immediate parents contributes on a decreasing scale to the make-up of the final offspring.

This law of ancestral heritage tells us to what degree the progressive thinning-out or exhaustion of heritable effects is unavoidable and explains the gradual disappearance of variations left to go their own gait. This can be adduced as an argument against the idea of definite fixation or the permanence of chance variations operating independently from the direct influence of the environment; which argument has been used in controversy raging around natural selection by those opponents who do not accept its exclusive nature for gospel.

In contrast to the Galtonian theory, Mendel's results are invoked to show that, at least in certain cases, hereditary characters do not become erased, and do not proceed to disperse into thin air in the course of a number of generations until vanishing entirely. This is doubtless so in some instances and in connection with particular characters, but not in all by any means. Characters arising from scarcely perceptible variations, which in accordance with Darwinian principles would necessarily continue to show themselves progressively, are not suitable for genetic research and hybridization studies. In such investigations, we must employ progenitors distinguishable by well-marked characters, and this is one reason for certain reservations when seeking to generalize Mendelian conclusions.

For reasons given elsewhere, we cannot accept the principle of representative particles. The very idea of a concrete particle in place of an abstract concept is illogical, and for that reason proponents of the theory have never been able to offer any evidence on the basis of experiment or actual observation, nor are they likely ever to do so. The explanation for the independent appearance of characters and of all other facts of inheritance which at first sight appear so easily accounted for by the theory of particles must therefore be sought elsewhere.

Agreement is, nevertheless, general in saying that the form of an offspring will resemble that of its direct ancestors, omitting now to include stability of characters from generation to generation. Heredity consists of an always orderly transmission of chemical, structural, anatomical, physiological and mental features of a distinct nature in the biological sense, from which the characters of the individual and the sum total of individuals making up a species all have their origin.

Reproduced by permission of the publisher

THOMAS HUNT MORGAN 1866–1945

The Theory of the Gene, New Haven: Yale University Press, 1928.

We are now in a position to formulate the theory of the gene. *The theory states that the characters of the individual are referable to paired elements (genes) in the germinal material that are held together in a definite number of linkage groups; it states that the members of each pair of genes separate when the germ-cells mature in accordance with Mendel's first law, and in consequence each germ-cell comes to contain one set only; it states that the members belonging to different linkage groups assort independently in accordance with Mendel's second law; it states that an orderly interchange— crossing-over—also takes place, at times, between the elements in corresponding linkage groups; and it states that the frequency of crossing-over furnishes evidence of the linear order of the elements in each linkage group and of the relative position of the elements with respect to each other.*

These principles, which, taken together, I have ventured to call "The Theory of the Gene," enable us to handle problems of genetics on a strictly numerical basis, and allow us to predict, with a great deal of precision, what will occur in any given situation. In these respects the theory fulfils the requirements of a scientific theory in the fullest sense.

In the light of what has just been said we can give a reasonable explanation of the differences that follow when a mutant change involves a whole chromosome (or part of one) and when only a single gene is involved. The former change adds nothing intrinsically new to the situation. More or less of what is already present is involved in the change, and the effects are small in degree but involve a large number of characters. The latter change—mutation in a single gene—may also produce widespread and slight effects, but, in addition, it often happens that one part of the body is changed to a striking degree along with other changes less striking. This latter kind of change, as I have said, supplies materials favourable for genetic study; these have been widely utilized. Now it is these mutational changes that have occupied the forefront of genetic publication, and have given rise to a popular illusion that each such mutant character is the effect of only one gene, and by implication, to the fallacy, more insidious still, that each unit character has a single representative in the germ material. On the contrary, the study of embryology shows that every organ of the body is the end-result, the culmination of a long series of processes. A change that affects any step in the process may be expected often to effect a change in the end-result. It is the final visible effect that we see, not the point at which the effect was brought about. If, as we may readily suppose, very many steps are involved in the development of a single organ, and if each of these steps is affected by the action of a host of genes, there can be no single representative in the germ-plasm for any organ of the body, however small or trivial that organ may be. Suppose, for instance, to take perhaps an extreme case, all the genes are instrumental in producing each organ of the body. This may only mean that they all produce chemical substances essential for the normal course of development. If now one gene is changed so that it produces some substance different from that which it produced before, the end-result may be affected, and if the change affects one organ predominatingly it may appear that one gene alone has produced this effect. In a strictly causal sense this is true, but the effect is produced only in conjunction with all the other genes. In other words, they are all still contributing, as before, to the end-result, which is different in so far as one of them is different.

In this sense, then, each gene has a specific effect on one particular organ but this gene is by no means the sole representative of that organ, and it has also equally specific effects on other organs, and, in extreme cases, perhaps on all the organs or characters of the body.

Continuing our comparison, the effect of a change in a gene (which, if recessive, means a pair of like genes) frequently produces a more localized effect than a doubling or trebling of the genes already present, because a change in one gene is more likely to upset the established relation between all the genes than is an increase in the number of genes already present. By extension, this argument seems to mean that each gene has a specific effect on the course of development, and this is not inconsistent with the point of view urged above, that all the genes or many of them work together towards a definite and complicated end-product . . .

Evidence points to the correctness of the theory that the sex-linked genes are carried by the chromosomes. Our present understanding of the mechanism of sex-determination has come from two sources. Students of the cell have discovered the role played by certain chromosomes and students of genetics have gone further and have discovered important facts as to the role of the genes.

Two principal types of mechanism for sex-determination are known.

They both involve the same principle, although they may seem, at first, to be the converse of each other.

The first type may be called the insect-type, because in insects we have the best cytological and genetic evidence for this kind of sex-determining mechanism. The second type may be called the avian type, because in birds we now have both cytological and genetic evidence for this alternative mechanism. It is also present in moths.

In the insect type the female has two sex-chromosomes that are called X-chromosomes. When the eggs of the female ripen (that is, after each has given off its two polar bodies), the number of the chromosomes is reduced to one-half. Each ripe egg, then, contains one X and, in addition, one set of ordinary chromosomes. The male has one X-chromosome only. In some species this X has no mate, but in other species it has a mate that is called the Y-chromosome. At one of the maturation divisions the X and the Y pass to opposite poles. One daughter-cell gets the X and the other the Y. At the other maturation division each splits into daughter chromosomes. The outcome is four cells that later become spermatozoa; two contain an X-chromosome, two contain a Y-chromosome.

Any egg fertilized by an X-sperm gives rise to a female that has two X's. Any egg that is fertilized by a Y-sperm gives rise to a male.

In the other type of sex-mechanism, the avian-moth type, the male has two like sex-chromosomes that may be called ZZ. These separate at one of the two maturation divisions and each ripe sperm-cell comes to contain one Z. The female has one Z-chromosome and a W-chromosome. When the eggs mature, each egg is left with one or the other of these chromosomes. Half the eggs contain a Z- and half contain a W-chromosome. Any W-egg fertilized by a Z-sperm produces a female (WZ). Any Z-egg fertilized by a Z-sperm produces a male (ZZ).

One of the surprises of the year 1923 was the simultaneous announcement by several independent workers (Santos, Kihara and Ono, Winge, Correns, and Blackburn) that in some of the flowering plants with separate sexes a mechanism is present that follows the XX-XY type. Several years before these observations in flowering plants had been made, it had been shown by the Marchals that when the spores are formed in dioecious mosses—mosses that have separate male and female gametophytes (or sexual prothallia)—two of the spores derived from the same sporophyte mother-cell produce female gametophytes and the other two male gametophytes. These experiments are interesting in showing how artificial hermaphroditic individuals may be made from plants that normally have separate sexes by combining the two sets of elements. The results also show that the sequence in which the sexual organs develop is determined by the age of the plant. More important is the actual reversal of this time relation by changing the genetic composition in the opposite direction . . .

If the same number of genes is present in a white blood corpuscle as in all the other cells of the body that constitutes a mammal, and if the former makes only an amoeba-like cell and the rest collectively a man, it scarcely seems necessary to postulate fewer genes for an amoeba or more for a man.

The only practical interest that a discussion of the question as to whether genes are organic molecules might have would relate to the nature of their stability. By stability, we might mean only that the gene tends to vary about a definite mode, or we might mean that the gene is stable in the sense that an organic molecule is stable. The genetic problem would be simplified if we could establish the latter interpretation. If, on the other hand, the gene is regarded as merely a quantity of so much material, we can give no satisfactory answer as to why it remains so constant through all the vicissitudes of out-crossing, unless we appeal to the mysterious powers of organization outside the genes that keep them constant. There is little hope at present of settling the question. A few years ago I attempted to make a calculation as to the size of the gene in the hope that it might throw a little light on the problem, but at present we lack sufficiently exact measurements to make such a calculation more than a speculation. It seemed to show that the order of magnitude of the gene is near that of the larger-sized organic molecules. If any weight can be attached to the result, it indicates, perhaps, that the gene is not too large for it to be considered as a chemical molecule, but further than this we are not justified in going. The gene might even then not be a molecule but only a collection of organic matter not held together in chemical combination.

When all this is given due weight it nevertheless is difficult to resist the fascinating assumption that the gene is constant because it represents an organic chemical entity. This is the simplest assumption that one can make at present, and since this view is consistent with all that is known about the stability of the gene it seems, at least, a good working hypothesis.

CONRAD HAL WADDINGTON *b.* 1905

An Introduction to Modern Genetics, New York: Macmillan, 1939.

THE CHROMOSOME AND THE GENE

Chromosomes as such have never been chemically analysed; they are too small for present methods. The nearest material which can be collected in quantities large enough for ordinary chemical investigation is sperm, particularly fish sperm. The head part of the sperm consists almost entirely of nuclear material, and analyses of this show that the two main constituents are thymonucleic acid (about 60 per cent) and simple proteins of the kind known as protamines (35 per cent)—Caspersson and Hammarsten, 1935. Nucleic acid combines very easily with proteins to form complex nucleoproteins, and it probably occurs in this combined form in the nucleus.

The distribution of nucleic acid in the nucleus can be investigated by means of ultra-violet spectroscopy (Caspersson, 1936), since it has a characteristic strong absorption at wave-lengths near 2,600 Ångströms. During division stages, when the chromosomes are contracted and can be seen as separate bodies, almost all the nucleic acid is attached to the chromosomes; it may also be in the chromosomes in the resting stages, but the fully extended chromonemata are not separately distinguishable from the nuclear sap, and the evidence is therefore not clear. The metaphase chromosomes can also be shown to contain protein, since they are attacked by proteolytic enzymes. In the salivary gland nuclei, the chromosomes contain the same two constituents, which perhaps makes it likely that the resting stage chromosomes are built up in the same way, and that the chromosome constitution is constant throughout the division cycle.

The chemical make-up of protein is not yet fully understood. It is known that some of the chemically rather inert proteins such as hair and silk are formed from fibrous elements consisting of chains of "polypeptide links," each link having the constitution -CO-CHR-NH-, where R is a group (the side-chain) which may be a simple hydrocarbon, an alcohol, or a base such as arginine. In the fibrous proteins we have mentioned, the links are arranged in linear chains, the chains being connected together by means of the side-chains. There are, however, other types of proteins in which the molecules seem to be spherical rather than elongated; these are known as the globular proteins, and there are others intermediate between the fibrous and globular types. We do not know to which of these types the chromosome proteins belong, since the protein isolated from sperm (clupein) has not been examined from this point of view. The thread-like appearance of the extended chromosome, and the two-ended nature of the chromomeres suggests that the chromosome consists of protein fibres arranged more or less parallel to its length. But this does not by any means necessitate the assumption that the chromosome protein is itself fibrous, since the orders of magnitude are quite different. The polypeptide links in a fibrous protein chain are about 0.35 mμ long by 0.45 mμ thick by 1 mμ wide in the direction of the side-chains. The visible chromonema is a few hundred millimicrons thick, while Müller's estimate, based on its length in salivary gland chromosomes, gives it a width of about 20 mμ. Fibres as large as this can just as well be formed from globular as from strictly fibrous proteins, since the units (molecules or repeat cells in crystals) of the former are about 6 mμ in diameter, and cases are known, for instance in some of the virus proteins, in which these units unite to form fibres a few tens of millimicrons thick.

Studies on the extensibility, and particularly the reversible extensibility, of the chromosomes give some, and could probably give much more, information about globular or fibrous nature of the chromosome proteins. In completely fibrous proteins the polypeptide chains lie fairly parallel and are more or less unfolded; they can only be stretched by actual straining of the chemical bonds. It is probable, however, that in globular proteins, the same or very similar polypeptide chains exist in a folded configuration, so that extension of a fibre constructed of globular protein involves only the unfolding of the chains and can proceed much farther before the fibre is ruptured. Duryee (1938) has shown that the lampbrush chromosomes of amphibian oocytes can be reversibly extended to about $3\frac{1}{2}$ times their normal length, at least under favourable conditions (in absence of calcium or other heavy metallic ions). Salivary gland chromosomes easily stretch to at least twice their length, and probably can be stretched farther when special efforts are made to do so by micro-dissection methods. Thus even in chromosomes in which the chromosome thread or chromonema is apparently uncoiled, the thread itself has considerable elasticity, and may perhaps be constructed of globular proteins in which the polypeptide chains are folded on a molecular scale. Much further study is required, however, before this can be taken as more than a suggestion.

When we turn to consider the other main constituent of chromosomes, the nucleic acid, a series of facts emerge which are extremely suggestive of an essential connection between nucleic acid and proteins, but whose exact significance cannot yet be stated. Nucleic acid itself easily forms fibres, and X-ray studies (Astbury and Bell, 1938) have shown that these consist of a chain of phosphoric acid residues to the side of which are attached a series of flat, plate-shaped groups each of which contains a purine base attached to a sugar. The first remarkable fact is that the repeat distance along the chain, i.e. the distance between neighbouring phosphoric acid residues, is almost exactly the same as the repeat distance in a polypeptide chain: $0.336 \, m\mu$ for the nucleic acid, $0.334 \, m\mu$ for the polypeptide. The difference, which may not be significant, is at least so small that it is easy to imagine that the polypeptide and nucleic acid chains might unite parallel to one another to give protein-nucleate chains. This can in fact actually be observed; Astbury has prepared the nucleate of clupein, the protein isolated from fish sperm, and shown that it is a fibrous material. Further confirmation comes from a study of the double refraction. Protein fibres have a somewhat weak double refraction which is positive in the direction of the fibre, while nucleic acid, in which there are large flat plate-like groups sticking out at right angles to the length of the fibres, has a much stronger double refraction which is negative in the direction of the fibre axis. The clupein-nucleate shows a double refraction negative in the fibre direction due to the nucleic acid. So do fully uncoiled chromosomes (Kuwada and Nakamura, 1934; Nakamura, 1937; Schmidt, 1936, 1937) such as those of salivary glands and zygotene stages; when the chromonema is presumably coiled in a single coil (e.g. mitotic metaphase), so that it runs perpendicularly to the length of the chromosome, the sign of the double refraction changes and becomes positive in the direction of the *chromosome* axis, while in meiotic metaphase, where the minor spiral is coiled again in a major spiral, the double refraction reverses again and becomes once more negative in the direction of the chromosome axis. All these data fit in very well with the idea that the protein and nucleic acid have combined to form composite fibres in which the two constituent fibres lie parallel to each other.

The cytological evidence makes it quite clear that the chromosomes are not homogeneous structures. In the first place, there is a differentiation in salivary gland chromosomes between the darkly staining bands and the non-staining inter-band regions. The property of stainability depends on the content of nucleic acid, and the concentration of this substance in the bands can be demonstrated directly by studies of ultra-violet absorption. One must suppose that the proteins in the band regions have a particular affinity for nucleic acid, and Wrinch (1936) has suggested that this may be due to a higher concentration of basic groups, particularly arginine, which is known to be present in remarkably high amounts in clupein. The difference between the bands and inter-bands appears, however, rather larger and more sharply defined than would be expected if it were due to a merely quantitative difference; but this appearance may turn out to be illusory when actual measurements of nucleic acid content become available.

The extensibility of the bands seems to differ sharply from that of the inter-bands, the former being much the more rigid. There are two factors to be taken into consideration here. Firstly, the nucleic acid fibre itself appears to be inelastic, and the rigidity of the bands may be due simply to their nucleic acid content. Secondly, while it is easy to see how nucleic acid may combine with fully extended polypeptides, it is not so clear how it can fit on to a globular protein; it is possible then that the proteins of the bands, when combined with nucleic acid, are in the extended form, and thus have themselves lost much of their extensibility; but whether this should be regarded as a result or as a contributory cause of their affinity for nucleic acid is as yet quite unknown.

The differential staining behaviour of the heterochromatic regions presumably depends on a chemical composition or physical state different to that of the euchromatin; but in general very little is known about this. In salivary glands of Drosophila, the inert regions (Bridges 1935, 1938; Prokofieva, 1935) show a structure of longitudinal striations and transverse bands which is somewhat similar to that of the euchromatic regions, except that the bands are more feebly staining and the whole structure less clear-cut. Although the inert region of the X chromosome, for instance, is fairly short in the salivary gland chromosomes and has only a small number of bands, it occupies a large proportion of the whole chromosome at metaphase of mitosis. This may perhaps be partly due to a lesser degree of spiralization in mitosis, although, since the region at that time is definitely shorter than it is in salivary chromosomes, some spiralization must occur. It is probable that the large relative volume of the inert region in mitosis is at least partly due to an abnormally large concentration of nucleic acid on to it at this stage. Muller (1938) claims that the greater bulk of the mitotic region is produced under the influence of only two loci in the region, and it is conceivable that the region contains loci specially concerned with the synthesis of nucleic acid.

The physical and chemical basis of this structure is unknown, but it is remarkable to find that the conditions underlying it appear to be transmissible; when inert regions are brought, by translocation, into contact with euchromatic parts of the chromosomes, there is a tendency for the latter to be modified in their appearance in salivary glands, so as to assume more nearly the inert structure (Prokofieva, 1935; Schultz and Caspersson, 1938). This suggests that the inert regions differ from the euchromatic regions only in some general condition which overlies the same basic differentiation into band and inter-band.

The centromeres are probably quite differently constituted from the rest of the chromosome. They seem to be unable to transmit torsional stresses, since the directions of coiling at metaphase are apparently independent of one another in the two arms of a chromosome with a central centromere; and similarly interference in crossing-over does not extend across a centromere. It has also been shown that when centromeres divide at metaphase they do not always split along a plane parallel to the length of the chromosome, but may occasionally be divided transversely or at any angle (Upcott, 1938). All these facts tend to suggest that the centromeres, unlike the rest of the chromosome, are not fibrous structures.

THE NATURE OF THE GENE

Before we can discuss the chemical nature of the gene, we must re-consider the definition of the word (Darlington, 1938). The Mendelian factor was originally defined simply as an entity which obeyed Mendel's laws and had an action in determining the characters of the adult organism. This definition could apply to whole chromosomes, or large sections of chromosomes. With the discovery of linkage and crossing-over, the definition of the factor, or, as it was now called, the gene, became narrowed down by adding the property that genes act as units in crossing-over, which occurs between them but not through them. At the present time, this definition has become unsatisfactory because there are cases to which it is inapplicable.

We know that genes may be guarded from crossing-over, for instance in the parts of the *Y* chromosome which have no homologue or in the complexes in a ring-forming heterozygote such as Oenothera. Moreover, we know of inert sections of the chromosomes, which appear to consist of particles much like those in the other regions of the chromosome, sharing with them the properties of multiplication and attraction, but lacking any effect on development; we may wish to stretch our definition to cover inert genes. Finally, the position effect shows that genes which may cross-over independently of one another may not be independent in their developmental actions; in this case the two parts of the definition do not tally.

It is clear that the old picture of the chromosome, as a linear array of individual indivisible particles, each of which is a gene, is too simple. In attempting to work out a more adequate picture, one can start from the fundamental fact that the chromosome is an elongated structure which, whenever we can analyse it, has differences arranged in a linear order along it; these differences can be detected by linkage studies, chromosome structures, etc. The units, between which differences are noted, may be of different sizes according to the different methods of investigation; there are, in roughly descending order, inert or pre-cociously condensing regions, large chromomeres, ultimate chromomeres or salivary gland chromomeres, and the units of cross-over and X-ray breakage. One might symbolically represent the chromosome thus: abcd′ e′f′g′hijklMNOPQ RSTU′V′W′, where there are differences on three scales, between the capitals and the lower-case letters; normal, underlined and dashed letters; and finally the letters themselves. The smallest units of this scheme, symbolized by the individual letters, are the units of crossing-over and X-ray breakage, and probably measure, as we have seen, about 100 mμ in length.

If we view the chromosome as it were through the other end of the telescope, attempting to build it up from chemical units, we arrive at a somewhat similar scheme of a linear order of units of different orders of magnitude. The ultimate units now are the links in a polypeptide chain, with a length of only 0·334 mμ. Exactly what the larger units are is more doubtful, but we have a range of possibilities; there are the periodicities along the chains (Bergmann and Niemann, 1938), the repeat units out of which protein crystals are built, the protein molecules such as they exist in solution, and finally virus particles, all of which may be considered as providing suggestions as to the kinds of units which may be involved. These units range in size nearly up to the 100 mμ which we took as an estimate of the smallest units to be considered when we approached the chromosome structure from the other end. It is, then, possible to conceive of the chromosome as a linear array of units, the units themselves forming a hierarchy all the way from heterochromatic and euchromatic regions, some tens of thousands of millimicrons long, to polypeptide links only a few tenths of a millimicron long.

This apparent homogeneity in the type of formal order exemplified by the chromosome on different scales should not tempt one to suppose that other properties may be just as easily conceived of in any of these scales. For instance, it is sometimes suggested that because the nature of one link in a polypeptide chain may chemically affect the properties of a neighbouring link, the same type mechanism may explain the phenomenon of position effect. But in the latter case, the influence is between neighbouring genes (i.e. breakage units) and extends over distances about a hundred times as great as in the former case. No direct analogy between mechanisms of two phenomena is possible; and in fact no example of a direct chemical influence extending throughout such a distance appears to be known in protein chemistry.

TABLE OF SIZES

The sizes are given in millimicrons (1 mμ = 10 Å = 10^{-6} mm). Where only one dimension is given, it is the diameter of a spherical unit.

Vaccinia virus	175
Rous sarcoma virus	100
Tobacco mosaic virus	430 $\times$ 12·3 $\times$ 12·3
Bushy stunt virus	28
Haemocyanin molecule	59 $\times$ 13·2 $\times$ 13·2
S13 Bacteriophage	10 (? shape)
Repeat unit of virus crystal	15 $\times$ 15 $\times$ 7
Haemoglobin molecule	2·8 $\times$ 0·6 $\times$ 0·6
Protein fibre (repeat unit)	0·334 $\times$ 0·45 $\times$ 1·0
Nucleic acid (repeat along fibre)	0·336

Gene (estimated maximum dimensions)	100 $\times$ 20 $\times$ 20
Sensitive volumes:	
Gene mutations (Timofeeff-Ressovsky)	*c.* 1
Gene (somatic, Haskins and Enzmann)	15
Cytological effects (Marschak)	5

Certain of the properties of the genes give some hints as to the possible kind of units which may fill the gap between the 0·334 mμ poly-peptide links and the 100 mμ genes. The most important is the property of identical reproduction. Between two cell divisions, each gene causes the formation of another gene exactly like it; if the gene mutates into an abnormal form, it is the mutated gene which is reduplicated. The gene, then, must in some way act as a model on which the new gene is formed. This can only occur if chemical forces originating in the radicals in the gene can extend far enough to influence the nature of radicals formed in the equivalent places in the new gene. The thickness which we can postulate for the gene is therefore limited by the distance through which we can imagine such chemical forces extending. Probably the maximum estimate which is chemically reasonable is about 10 mμ, which is the order of magnitude of the thickness of the repeat units out of which protein crystals are built. This is of the same order of magnitude as the estimate given above for the maximum thickness of the chromosome thread. It is therefore impossible to reject, from consideration of gene reduplication, the idea that the gene is a single unit. On the other hand, a further difficulty arises in this connection, namely the necessity to find some mechanism which accounts for the fact that only two genes, the old one and the new, are present at the end of each intermitotic period. The reduplication occurs only once. No plausible hypothesis to account for this has been put forward.

Alternatively, we may assume that the gene is compound, consisting of a number of identical sub-units. Such a supposition probably simplifies the task of accounting for gene reproduction. The chemical forces on which the identity of the new and old gene depend would not have to extend so far from the radicals to which they were due, since the thickness of the sub-units would be less than that estimated for the whole chromosome thread. Similarly the reproduction might continue gradually, and the gene grow until it eventually split into two by reason of some instability which increased with increasing size, such as that which causes a drop to break up when it passes a certain size limit. The difficulty of this hypothesis, as was pointed out before, is the fact that some genes (though only a few) show more or less equal rates of back and forward mutation.

It appears not unlikely that nucleic acid plays some important role in the process of gene reduplication. For instance, the most rapid synthesis of nucleic acid occurs just before the prophase of mitosis, at the time when the chromosome appears to split or reduplicate. Again, it is remarkable that the virus proteins, which share with the genes the property of identical reproduction in living systems (Stanley, 1938), and of mutation also (McKinnery, 1937), contain large quantities of nucleic acid. Conceivably there is some connection here with the remarkable fact recently revealed by Schultz and Caspersson (1938), that nucleic acid is in some way connected with the stability of the gene; when parts of the inert region in Drosophila are translocated into the euchromatic regions, they frequently cause the neighbouring loci to become unstable and undergo somatic mutations which give rise to phenotypic spotting such as that found with other mutable genes; and this instability appears to be correlated with an increase in the nucleic acid content of the corresponding bands in the salivary gland chromosomes.

All the above considerations apply to genes considered as units of crossing-over and X-ray breakage. It is quite possible that only a small part of the gene defined in this way is actually active in the control of development. We cannot rule out the possibility that this activity is due to some particular group within the large protein-nucleic acid complex we have been discussing. In fact, the small size of the "sensitive volumes" found for particular steps of mutation might suggest that only quite restricted regions are concerned in producing the phenotypic differences between two allelomorphs; but we have pointed out the many uncertainties in the interpretation of the sensitive volume measurements.

On the other hand, it is quite possible that all primary gene products are enzymes and therefore probably proteins, which may be similar in composition to the genes themselves. It would then be in order to suggest a connection between gene activity, in which enzymes were produced and liberated into the cytoplasm, and gene reproduction, in which similar bodies were formed but retained in the neighbourhood to form a new chromosome.

It will be apparent from the above discussion that the exact knowledge at our disposal is so meagre that very many alternative hypotheses are still possible as to the nature of the chromosome, and the gene in its different senses. However, the enormously important effects of the genes on development, their capacity for identical reproduction, and the fact that they, rather than the cells of an earlier time, seem to be the most ultimate units into which we can analyse living organisms, make the problem of their constitution one of the most fundamental questions of biochemistry, well worthy of discussion even long before it can be fully answered.

CHAPTER X

INDIVIDUAL AND SPECIES

THE INDIVIDUAL

An individual can be an animal or a plant. Common observation shows that the definition of the word individual should be simple, e.g. dog, shark, rose-bush, beech-tree, etc. Nevertheless, sometimes insuperable doubts arise when we try to define what individuality is. Since all living organisms are formed from cells, since there are some plants and animals in existence composed of one cell only, whilst others are combinations of several different cells generally great in number, it is at times difficult if not impossible to distinguish a colony of several single-celled individuals from a polyplastid. We have already alluded to this point.

Hertwig in 1894 defined the individual organism as a living unit capable of preserving its shape, endowed with the general functions of existence, and able to retain its stability or constancy despite variations in the surrounding medium. Individuality may then be defined by its power of *making, preserving, rebuilding and reproducing* a particular shape or form. It is all the same whether we are dealing with a single-celled organism or one of the more outstanding plants or creatures in this world. A part of an animal or of a plant which goes on living for a period after it has been detached from the main body will not become a fresh individual unless it is able to *reconstruct* a complete animal or plant identical to the one from which it had its origin.

The individual will therefore exist and produce fresh individuals by reproduction, whatever may be the process by which such reproduction is effected, and will follow a course in time, which is life in its various succeeding stages all exactly mapped out until the end comes in death. Individuality may be said therefore, from the morphological and functional standpoint, to be a complete whole in itself, which in the case of mankind is subjectively translated into consciousness, feeling, and knowledge of the ego. It is through anatomy, physiology and psychology that the individual becomes apart from the world, and opposes himself to his surroundings, forming thus an independent entity. Withal, existence of an individual without a surrounding medium would be impossible.

SPECIES

An aggregate of individuals, animals or plants, which present broad similarities without being absolutely identical, forms the species. The source of such resemblances lies in the fact that individuals of the same species reproduce and give birth to others like their progenitors. Cuvier in 1817 defined species as a collection of individuals descending one from the other having their source in some common parentage and resembling therefore each other as much as they resemble such original parents.

A purely morphological criterion cannot be applied in all cases, since, even within a single species, differences of form are to be found, for instance: according to sex; to the period of existence— larva or imago; to conditions in the environment, and so forth. It would be more precise to attribute the characteristics of individuality to the chemical composition and physico-chemical structure of the living plasma. Individuals of the same species will be therefore those whose cells and humours present similar chemical and physico-chemical features, from which the functional and morphological characteristics are precisely derivable.

RACE OR VARIETY

A species includes races and varieties. Racial differences may perhaps lie so deep that it is not always easy to distinguish a specific character from a racial one; in certain cases two races may from their form appear to be more remote from each other than two species. Here too the morphological criterion is not sufficient; only the properties of plasm and tissue will allow us to define a species and distinguish it from a variety. As a consequence of this the safest character for differentiating is the capacity for reproduction and fertility. In general *half-breeds* are distinguished from *hybrids*. The increasing affinity of living plasma as relationship becomes closer is one explanation of the facts observed when we try to reproduce different breeds. There is no method of crossing two sufficiently remote species, though, on the contrary, we certainly can and do get hybrids of species fairly close to each other. Such hybrids are usually sterile, either in the immediate offspring or petering out after a few generations. The crossing of animals and plants of the same species, and of different race, gives rise to half-breeds which are just as fruitful as pure stock. These differences of reproductive capacity are due to the greater or lesser identity of cell plasm in the various cases and constitute a further important argument in favour of the theory that the chemical and physico-chemical quality of the living elements is the essential characteristic of a species.

BALANCE AND SPECIES

Species represents a biological balance, and an individual is also a balance or system in equilibrium. Within a species itself, individuals are similar but not identical. The differences are called *variations*, and when numbers of individuals exhibit the same variation they form a variety. A variety is distinguished from a race because the latter has greater constancy, whereas the former fluctuates in general and tends to disappear, whilst individuals showing such variation engender a line which continues over a larger or smaller number of generations and becomes approximated to the specific type. This is what Galton in 1889 termed "filial regression."

Specific characters do exist representing a middle term, and on one side or the other of this average, we find variations which arithmetically follow Gauss's Law, or the Law of Mean Errors of the Calculus of Probabilities. This can be confirmed by taking the more distinctive characters. Thus height, weight, colour of eyes, basic intensity of metabolism, pulse, blood pressure, and so on. In species other than the human we can take for instance the weight of certain seeds, the height of a plant, the number of vertebrae in certain fish, and so forth. The graph showing such variations quantitatively is called a "Galton's curve," and demonstrates that individuals showing typical average characters of the species are the most numerous, whilst divergencies above and below such average diminish in proportion to the degree of variation of the character from the average type. Thus, for instance, the height of individuals: there is an average height, a distinctive feature for each species and this will be the height of the majority of specimens. There are taller individuals and shorter ones, and the numbers of these outsiders decreases as their height differs more and more from the standard.

There is furthermore, as we have repeatedly mentioned, a tendency to revert to the specific standard. Short parents and tall usually have offspring whose height is different from their own and more or less in accordance with the variance of the parents from the average standard. In all these cases the tendency for "reversion to type," to "balance" in the species, is evident.

ELEMENTAL SPECIES

There may well exist primitive species with a certain degree of stability, fairly constant in form, such as those distinguished by Jordan in 1848 as between the plants and the midpoint of the limits of a Linnaean species. Thus, for instance, in *Viola tricolor*, Jordan distinguished several dozen various species, independent and stable as regards the transmission of their characters through the seeds: with large, small and middling blossoms respectively identifiable through particular features of the component parts of the blossoms, seeds, and so on. The same thing occurs in many other plants, and animals too, though the latter usually have greater stability. It is consequently often difficult to make a distinction between this idea of elemental species and that of variety and race. The most rigorous test is founded on the possibility of hereditary transmission previously pointed out. A variety tends to disappear in the course of a few generations; race is more constant, whilst species is still more so.

The researches of Ewing on greenfly, of Johannsen on kidney beans, of Zeleny, Mattoon and Nay on the fruit-fly *Drosophila*, of MacDowell, Payen, de Hoge and others, using in some cases parthenogenetic stock and in others taking observations on the effects of sexual reproduction, all lead to the same conclusion, namely that the genes are invariable and heredity is homogeneous. If individuals be submitted, through a series of generations, to outside influences which provoke variations in them, then in so far as such influences cease to act or lose their power, the species will after a few generations revert to its normal type. Daphnia, for instance, kept at a high temperature alter the shape of their head, and such variation may pass on to the next generation, but when the modifying agent ceases to apply, the very next subsequent generation will come out with heads of normal shape and size. There are many other examples of a similar nature.

THE EVOLUTION OF SPECIES

Variations in the species which populated the earth during past geological periods do not seem to corroborate such stability. Palaeontology shows that in ancient times species existed which have since disappeared, whilst conversely we now have species which were not formerly in existence, if we accept as evidence the absence of their remains among fossils. Fauna and flora have changed in the course of time, generally for the better.

This led to the idea (Nägeli, 1865–84; Eimer, 1888) that species might have evolved by orthogeny, with variations accumulating from generation to generation and diverging thus ever increasingly from the primal type by a kind of differentiation which continuously made itself evident by setting in motion a "principle of perfection," or some "internal force" or "law of progress," that is to say a phylogenetic differentiation comparable to the differentiation which comes about during the development of the embryo, and which is similarly subject to internal forces intrinsic in the properties of organic matter, though itself lying totally outside our present knowledge.

It is not to be wondered at that phylogenetic development has many a time and oft been compared to that of the embryo. As far back as 1821, Meckel asserted that there was some similarity between the formation of an embryo and the various forms of species in the animal kingdom. Von Baer in 1828 noted that in ontogeny the higher vertebrates pass through stages at which the lower ones are arrested: "The embryos of mammals, birds, lizards and serpents are extraordinarily alike during the early stages of their development, in both the general and the particular, whence it is difficult to distinguish one from the other." Thus it is that Häckel in 1866 came to propound his *fundamental law of biogenesis*: "The organized individual repeats during the rapid course of its own development the more important phases through which its ancestors passed in the course of their long palaeontological evolution." In other words, "Ontogeny may be said to be a conspectus of phylogeny."

From the instant that orthogenetic evolution is not ascertainable and there is no further assurance that the species may tend to become differentiated by divergence, it is no longer feasible to make a comparison between phylogeny and ontogeny, where indeed there is a progressive multiplicative and differentiable evolution starting from the zygote. It must, however, be conceded that fossil plant and animal forms certainly show greater and greater complexity and a more evident differentiation in their organs the closer the period of their existence approaches the present time.

ADAPTATION TO ENVIRONMENT

The theory of orthogeny is opposed by that of adaptation to environment. Lamarck in 1809 founded his philosophy of the animal kingdom mainly on the idea of adaptation. Living bodies assume their distinguishing forms and characters through the influence of the medium in which they exist; and the characters thus acquired are transmitted by inheritance, whereby such forms continue evolving in accordance with the needs of existence. The species is thus the result of a conflict between the living body and its surroundings and changes in accordance with such surroundings.

Lamarck's ideas have been bitterly assailed by those biologists who do not agree that acquired characters can be transmitted by inheritance; but, in spite of all objections, these notions still continue to exercise wide influence on our present framework of ideas, which cannot by any means rest satisfied within the narrow limits of the Weismann system.

It is not in fact reasonable to doubt that some correlation between the medium and the living creature does exist and that the state of the latter is conditional upon that of its environment, for unless this were so life would become impossible. We shall deal with this point in greater detail later on. The influence of environment is shown with special effect by the appearance and development of suitable forms and shapes in animals and plants, thus enabling the continuance of existence in a particular medium.

MUTATION

Another way of changing the forms of living creatures, which could give us some explanation of how new species came to swell the number of those already in existence, is mutation or sudden variation. From time immemorial, practical botanists, zoologists and naturalists have been aware of what Darwin eventually labelled "sports." Waagen in 1849 suggested the name "mutation," and Kölliker in 1864 and Dall in 1867 made certain researches in this direction. It was not however until 1901 that de Vries gave an exact description and noted the great significance of this phenomenon. He describes the case of the *Oenothera Lamarckiana* which produces monster shrubs *Oenothera gigas*, apparently spontaneously and without any valid reason, alongside normal plants. There are also other types recognizable from the shape of their leaves, and others from different characters. The mutate form can be transmitted through inheritance.

At first it was thought, on the basis of de Vries's discovery, that mutation was the exception rather than the rule and that at a particular moment a form might arise different from the specific, but this only occasionally. It was seen later that *Oenothera* is by no means a unique case, quite the contrary. Thus de Vries writes: "Species are not continuous, they exhibit sudden changes and leaps, to new types representing a species distinct from the one whence they proceed. Such new species appear without warning and without any intermediate transition stage."

After the discovery of mutations, it was possible to prove that the matter was not one of exceptional fact. It is indeed a normal and very frequent process of modification. In *Drosophila* alone, among two and a half million fruit-flies studied by Morgan and his co-workers, practically a thousand forms of mutation were counted. It is understandable then that as fresh characters are transmitted by inheritance, mutation may well constitute a starting-point for variations of species perhaps of considerable importance. Such variations will sometimes be temporary, sometimes permanent.

The main interest in mutations rests in the fact that it has been possible to establish the relation thereof with modifications in the chromosomes or genes. In this way, the study of mutations has been of inestimable value in the field of genetics, leading for instance to knowledge of the *multiple alleles* which is of far-reaching importance in human pathology, and to the experimental production of artificial mutates. These experimental cases which at first were thought to be always spontaneous in origin and without appreciable cause can nowadays be directed in such a manner as to produce fresh forms of species.

Müller in 1927 produced mutations for the first time by means of radium beta-rays. The flow of electrons increased the mutation process by as much as 150 to 1. Mutations can also be obtained by means of X-rays, also through changes of temperature. The metabolic changes which they entail also foster mutation. Indeed much influence must be attributed to the metabolic condition of the subject and to chemical factors which have yet to be determined, as well as modifications within the body and in the medium. Blaringhem obtained mutations by causing violent mutilations in Indian corn stalks. Mutations are more frequent among plants than among animals and still rarer the higher the creature's place may be in the animal kingdom. We must not forget, however, that even among the mammals there are typical cases of mutation. For instance, the short-legged sheep of Panama first bred in 1791, horn-less Herefords in Kansas in 1889, tail-less or fur-free mousers and ratters both canine and feline, and sharp modifications of character in horses, rats and other animals.

Mutation cases correspond essentially to what Lloyd Morgan in 1928 termed "emergences," the appearance of fresh properties thanks to fresh combinations of factors. It is said, for instance, that if a person had no knowledge of water but did on the other hand know oxygen and hydrogen, he would never be able to foretell that the combination of the two gases would result in a compound like water with properties so different from those of the original two simple elements, nor would it be possible to foresee the vital properties of living matter resulting from complex chemical combinations. Changes in composition and structure of molecules might give rise to the formation of live matter and to the special characteristics thereof.

NATURAL SELECTION

Natural selection operates consequently on differences of species occurring through orthogenesis, adaptation or mutation. By its effects the unfit disappear and the fit survive, so that selection determines the species which inhabit each of the various regions on this earth.

Natural selection is a term applicable to two distinct ideas, namely the Darwinian "struggle for life," or the mere "effort to exist" which was Darwin's original thought; or alternatively a violent conflict according to the interpretation put on it by so many, which in plain words becomes the result of a battle of the living creature against its physical and biological surroundings.

Nobody can deny that environment determines the beings which will survive in it. If, in reality, mechanisms exist which cause the appearance of fresh species in the course of time, some of these species will disappear because they cannot withstand the surrounding conditions, whilst others will persist, being of such nature that the medium will be favourable to them. This is an incontestable fact, whatever one's personal opinion may be with regard to the origin of species.

Natural selection has since Darwin's time come to be considered as the certain means of originating specific forms. Weismann, and with him many others, considered that natural selection would resolve all difficulties when seeking to explain the origin of species. The historic controversy between Weismann and Spencer on this question will long be remembered. Selection is not a metaphysical principle or a driving force, it is merely a biological fact to which a disproportionate theoretical significance has been given.

Species, as we have seen, represents a position of balance, an equilibrium between internal factors among themselves and between them and outside influences. By this balance, which is one among the many which make life possible, certain living forms preserve, for a longer·or shorter period, characters which they transmit through heredity and which therefore endure from generation to generation. Huxley in 1880 stated that species is an abstraction, a somewhat strong statement assuredly, since species is an aggregation or total of real beings and individuals, and any aggregate contains the reality of its components. Species has its history, its life course, just as an individual has. A species may make its appearance, wax strong, decay and finally become extinct, because each individual varies in vitality, since the fruitfulness of such individuals changes, or because the power to withstand the medium alters, or by reason of various combinations of these factors. These are what cause a species to progress or to deteriorate. The problems relating to species, its origin and its conditions of existence are the most basic applying to biological science. As much attention must be devoted to consideration of the biology of species as to that of the individual. However, these serious problems must be denuded of anything which may not be of pure scientific value and interest, so as to avoid unjustifiable and irksome disputes.

* * * * *

CAROLUS LINNAEUS (CARL VON LINNÉ)
1707–78

Systema Genera, Species, Plantarum (Plant Genera and Species), Stockholm: Laurentius Salvii, 1753.

NOTE DATED 2ND MAY, 1753, AT UPSALA, SWEDEN

I am a Man of Science contemplating the World, a Workroom of the Almighty, embellished where'er one looks with Miracles produced by that Supreme All-knowing Power, a place of Delight when one has attained a true Knowledge of Our Lord. Though but a lowly guest among such great Endowments, I still know how to appreciate the Benefits thus conferred.

To render ourselves worthy of such Liberality, we must study with all due Reverence these Works of the Creator of all things, in so far as they may be understood by application of our Reason and our Human Capacities.

Knowledge implies Peculiar Ideas and suitable Designations, some of which allow us in ourselves to imitate certain things, and others which make possible an Intercourse with the rest of Mankind by means of some comprehensible Manner of Parlance.

The Natural Sciences surrounding Man include Physics and Chemistry dealing with the elements, whilst Zoology, Botany and Mineralogy cover the Study of the Natural Entities.

Botany, the Science of Plant Life, in its early Beginnings knew but few plants. Now, it constitutes a vast Science thanks to the Labours of countless Learned Men throughout the Ages. Despite this, the names given to Plants were determined by Caprice, for which reason its study was rendered difficult and the results obtained from it were but doubtful.

Learned Men investigated the Orders of Nature and built up this Science upon solid Foundations by establishing a proper System. Let us mention Gessner, Cesalpino, Bauhin, Hermann, Tournefort and Vaillant among countless others.

The Open Sesame to Method commences with Genus, divided into Species, a logical concept differentiating all things with Certainty. To distinguish the intrinsic Characters of plants, in order thus to find ourselves in a position to give them Specific Names is by no means an easy Task. It necessitates abundant and carefully gathered collections, painstaking examination of the separate parts, selection by differences, and a correct application of conclusions. Let us be on Guard against those who desecrate the Good Name of Botany, when they propound trifling denominations, lacking in specific Discrimination, which are liable to lead us back toward a primitive Crudeness in Science.

GEORGES LÉOPOLD CUVIER 1769–1832

Le Règne Animal, distribué d'après son Organisation (The Animal Kingdom arranged according to its Organization), Paris: Déterville, 1817.

Any investigation relating to natural beings needs some way of distinguishing each one of them in such a manner as to avoid confusion with the countless others which Dame Nature brings to our notice. Natural History ought therefore to possess as a basic tool some "Natural System" or comprehensive list containing an agreed name for every single being, thus enabling it to be identified by reference to its distinctive features.

As no living organism is entirely simple in its character, it is insufficient to try to distinguish it merely by one single feature of its make-up. We must therefore, nearly always, rely on a combination of several distinct features if we wish to differentiate any particular individual from its fellows, which latter may in part, if not entirely, through their very similarity show traits pertaining to the first specimen. In a like manner, such original sample may present characters common to itself as well as to all the others and will then be distinguishable only through those features not possessed by the remainder. Some beings are very closely related to one another in the biological sense and possess many characters in common, whilst others show far less a degree of similarity in detail. The larger the quantity of beings requiring identification, the greater the number of characters will it be necessary to take into consideration, even perhaps down to the point of making a complete and exhaustive specification of every single detail of the living form.

This state of affairs necessitates the setting-up of divisions and subdivisions. As mentioned above, some individuals will show a large number of features in common, and may then be grouped together as a *genus*, wherein differences in detail are slight. The genera will then have to be classified and, repeating the operation, closely related genera will be combined to form an *order*. Similarly, orders related in rank will form a *class*.

All this implies some framework and method, and nothing could be more suitable than the "Natural Method," that is to say, an arrangement showing us how individual members of one particular genus are more like each other than to members of other genera, and how the genera of a particular order are more alike to each other than to those of other orders, and so on. This is the system towards which Natural History must necessarily tend, since obviously if an exact system of classification can be attained, the complete portrayal of Nature will have been achieved. Every single living body can be distinguished from the rest by means of its similarities and differences, and if a classification like that mentioned above be possible, then the relationships of living bodies among themselves will become clearly demonstrable.

The Natural Method is the proper scientific way of doing things, and every step made in improving it will be a step forward towards the objectives of our Sciences.

Since existence itself is the most important of the qualities and the ultimate supreme characteristic of Nature, it is not surprising that throughout the ages, the most outstanding principle of distinction has been that leading to the classification of natural organisms into two huge divisions: the *living* and the *inorganic*.

CHARLES DARWIN 1809–82

The Origin of Species, London: Murray, 1859.

How will the struggle for existence, briefly discussed in the last chapter, act in regard to variation? Can the principle of selection, which we have seen is so potent in the hands of man, apply under nature? I think we shall see that it can act most efficiently. Let the endless number of slight variations and individual differences occurring in our domestic productions, and, in a lesser degree, in those under nature, be borne in mind; as well as the strength of the hereditary tendency. Under domestication, it may truly be said that the whole organization becomes in some degree plastic . . .

Let it also be borne in mind how infinitely complex and close-fitting are the mutual relations of all organic beings to each other and to their physical conditions of life; and consequently what infinitely varied diversities of structure might be of use to each being under changing conditions of life. Can it, then, be thought improbable, seeing that variations useful to man have undoubtedly occurred, that other variations useful in some way to each being in the great and complex battle of life, should occur in the course of many successive generations? If such do occur, can we doubt (remembering that many more individuals are born than can possibly survive) that individuals having any advantage, however slight, over others, would have the best chance of surviving and of procreating their kind? On the other hand, we may feel sure that any variation in the least degree injurious would be rigidly destroyed. This preservation of favourable individual differences and variations, and the destruction of those which are injurious, I have called Natural Selection, or the Survival of the Fittest. Variations neither useful nor injurious would not be affected by natural selection, and would be left either a fluctuating element, as perhaps we see in certain polymorphic species, or would ultimately become fixed, owing to the nature of the organism and the nature of the conditions . . .

We shall best understand the probable course of natural selection by taking the case of a country undergoing some slight physical change, for instance, of climate. The proportional numbers of its inhabitants will almost immediately undergo a change, and some species will probably become extinct. We may conclude from what we have seen of the intimate and complex manner in which the inhabitants of each country are bound together, that any change in the numerical proportions of the inhabitants, independently of the change of climate itself, would seriously affect the others. If the country were open on its borders, new forms would certainly immigrate and this would likewise seriously disturb the relations of some of the former inhabitants. Let it be remembered how powerful the influence of a single introduced tree or mammal has been shown to be. But in the case of an island, or of a country partly surrounded by barriers, into which new and better adapted forms could not freely enter, we should then have place in the economy of nature which would assuredly be better filled up, if some of the original inhabitants were in some manner modified; for, had the area been open to immigration, these same places would have been seized on by intruders. In such cases, slight modifications, which in any way favoured the individuals of any species, by better adapting them to their altered conditions, would tend to be preserved; and natural selection would have free scope for the work of improvement.

We have good reason to believe that changes in the conditions of life give a tendency to increased variability; and in the foregoing cases the conditions have changed, and this would manifestly be favourable to natural selection, by affording a better chance of the occurrence of profitable variations. Unless such occur, natural selection can do nothing. Under the term of "variations," it must never be forgotten that mere individual differences are included. As man can produce a great result with his domestic animals and plants by adding up in any given direction individual differences, so could natural selection, but far more easily, from having incomparably longer time for action. Nor do I believe that any great physical change, as of climate, or any unusual degree of isolation to check immigration, is necessary in order that new and unoccupied places should be left, for natural selection to fill up by improving some of the varying inhabitants. For as all the inhabitants of each country are struggling together with nicely balanced forces, extremely slight modifications in the structure or habits of one species would often give it an advantage over others; and still further modifications of the same kind would often still further increase the advantage so long as the species continued under the same conditions of life and profited by similar means of subsistence and defence. No country can be named in which all the native inhabitants are now so perfectly adapted to each other and to the physical conditions under which they live, that none of them could be still better adapted or improved; for in all countries, the natives have been so far conquered by naturalized productions, that they have allowed some foreigners to take firm possession of the land. And as foreigners have thus in every country beaten some of the natives, we may safely conclude that the natives might have been modified with advantage, so as to have better resisted the intruders.

As man can produce, and certainly has produced, a great result by his methodical and unconscious means of selection, what may not natural selection effect? Man can act only on external and visible characters: Nature, if I may be allowed to personify the natural preservation or survival of the fittest, cares nothing for appearances, except in so far as they are useful to any being. She can act on every internal organ, on every shade of constitutional difference, on the whole machinery of life. Man selects only for his own good: Nature only for that of the being which she tends . . .

It may metaphorically be said that natural selection is daily and hourly scrutinizing, throughout the world, the slightest variations; rejecting those that are bad, preserving and adding up all that are good; silently and insensibly working, *whenever and wherever opportunity offers*, at the improvement of each organic being in relation to its organic and inorganic conditions of life. We see nothing of these slow changes in progress, until the hand of time has marked the lapse of ages, and then so imperfect is our view into long-past geological ages that we see only that the forms of life are now different from what they formerly were.

In order that any great amount of modification should be effected in a species, a variety when once formed must again, perhaps after a long interval of time, vary or present individual differences of the same favourable nature as before, and these must be again preserved, and so onwards step by step. Seeing that individual differences of the same kind perpetually recur, this can hardly be considered as an unwarrantable assumption. But whether it is true, we can judge only by seeing how far the hypothesis accords with and explains the general phenomena of nature. On the other hand, the ordinary belief that the amount of possible variation is a strictly limited quantity is likewise a simple assumption.

THOMAS HENRY HUXLEY 1825–95

The Crayfish: An Introduction to the Study of Zoology, London: Kegan Paul, 1880.

All the individual crayfish referred to thus far, therefore, have been sorted out, first into the groups termed *species*; and then these species have been further sorted into two divisions termed *genera*. Each genus is an abstraction, formed by summing up the common characters of the species which it includes, just as each species is an abstraction composed of the common characters of the individuals which belong to it; and the one has no more existence in nature than the other. The definition of the genus is simply a statement of the plan of structure which is common to all the species included under that genus; just as the definition of the species is a statement of the common plan of structure which runs throughout the individuals which compose the species. The southern crayfishes, like those of the northern hemisphere, are divisible into many species; and these species are susceptible of being grouped into six genera: *Astacoides, Astacopsis, Chaeraps, Parastacus, Engaeus* and *Paranephrops* forming the Southern family *Parastacidae*, on the same principle as that which led to the grouping of the Northern forms into two genera *Astacus* and *Cambarus* of the family *Potamobiidae*, the two families forming the general tribe of *Prostacidae*.

The same convenience which has led to the association of groups of similar species into genera, has given rise to the combination of allied genera into higher groups, which are termed *families*. It is obvious that the definition of a family, as a statement of the characters in which certain number of genera agree, is another morphological abstraction, which stands in the same relation to generic, as generic do to specific abstractions. Moreover, the distinction of the family is a statement of the plan of all the genera comprised in that family, and carrying out the metaphorical nomenclature of the zoologist a stage further, we may say the families form a *tribe*—the definition of which describes the plan which is common to several families.

It may conduce to intelligibility if these results are put into a graphic form, the plan of an animal in which all the externally visible parts which are found, more or less modified, in the natural objects which we call individual crayfishes. *A* represents the plan of the tribe. *B* exhibits such a modification of *A* as converts it into the plan common to the whole family of *Parastacidae* (Southern crayfishes). *C* stands in the same relation to the *Potamobiidae* (Northern crayfishes). All these figures would represent abstractions—mental images which have no existence outside the mind.

The truths of anatomy and embryology are generalized statements of facts of experience; the question whether an animal is more or less like another in its structure and its development, or not, is capable of being tested by observation; the doctrine of the unity of organization of plants and animals is simply a mode of stating the conclusions drawn from experience. But, if it is a just mode of stating these conclusions, then it is undoubtedly conceivable that all plants and all animals may have been evolved from a common physical basis of life, by processes similar to those which we every day see at work in the evolution of individual animals and plants from that foundation.

HUGO DE VRIES 1848–1935

Species and Varieties: Their Origin by Mutation, London: Kegan Paul, 1905.

From all these statements and a good many others which can be found in horticultural and botanical literature, it may be inferred that mutations are not so very rare in nature. Moreover we may conclude that it is a general rule that they are neither preceded nor accompanied by intermediate steps, and that they are probably constant from seed from the first.

Why then are they not met with more often? In my opinion it is the struggle for life which is the cause of this apparent rarity; which is nothing else than the premature death of all the individuals that so vary from the common type of their species as to be incapable of development under prevailing circumstances. It is obviously without consequence whether these deviations are of a fluctuating or of a mutating nature. Hence we may conclude that useless mutations will soon die out and will disappear without leaving any progeny. Even if they are produced again and again by the same strain, but under the same unfavourable conditions, there will be no appreciable result.

Thousands of mutations may take place yearly among the plants of our immediate vicinity without any chance of being discovered.

We are trained to the appreciation of the differentiating marks of systematic species. When we have succeeded in discerning these as given by our local flora lists, we must rest content. Meeting them again we are in the habit of greeting them with their proper names. Such is the satisfaction ensuing from this knowledge that we do not feel any inclination for further inquiry. Striking deviations, such as many varietal characters, may be remarked, but then they are considered as being of only secondary interest. Our minds are turned from the delicately shaded features which differentiate elementary species.

Even in the native field of the evening-primroses, no botanist would have discovered the rosettes with smaller or paler leaves, constituting the first signs of the new species. Only by the guidance of a distinct theoretical idea were they discovered, and having once been pointed out a closer inspection soon disclosed their number.

Variability seems to us to be very general, but very limited. The limits, however, are distinctly drawn by the struggle for existence. Of course the chance for useful mutations is a very small one. We have seen that the same mutations are as a rule repeated from time to time by the same species. Now, if a useful mutation, or even a wholly indifferent one, might easily be produced, it would have been so, long ago, and would at the present time simply exist as a systematic variety. If produced anew somewhere the botanist would take it for the old variety and would omit to make any inquiry as to its local origin.

Thousands of seeds with perhaps wide circles of variability are ripened each year, but only those that belong to the existing old narrow circles survive. How different would Nature appear to us if she were free to evolve all her potentialities!

Darwin himself was struck with this lack of harmony between common observations and the probable real state of things. He discussed it in connection with the cranesbill of the Pyrenees (*Geranium pyrenaicum*). He described how this fine little plant, which has never been extensively cultivated, had escaped from a garden in Staffordshire and had succeeded in multiplying itself so as to occupy a large area. In doing so it had evidently found place for an uncommonly large number of plantlets from its seeds and correspondingly it had commenced to vary in almost all organs and qualities and nearly in all imaginable directions. It displayed under these exceptional circumstances a capacity which never had been exceeded and which of course would have remained concealed if its multiplication had been checked in the ordinary way . . .

From this discussion we may infer that the chances of discovering new mutating species are great enough to justify the utmost efforts to secure them. It is only necessary to observe large numbers of plants, grown under circumstances which allow the best opportunities for all the seeds. And as nature affords such opportunities only at rare intervals, we should make use of artificial methods. Large quantities of seed should be gathered from wild plants and sowed under very favourable conditions, giving all the nourishment and space required to the young seedlings. It is recommended that they be sown under glass, either in a glass-house or protected against cold and rain by glass-frames. The same lot of seed will be seen to yield twice or thrice as many seedlings if thus protected, compared with what it would have produced when sown in the field or in the garden. I have nearly wholly given up sowing seeds in my garden, as circumstances can be controlled and determined with greater exactitude when the sowing is done in a glass-house.

The best proof perhaps, of the unfavourable influence of external conditions for slightly deteriorated deviations is afforded by variegated leaves. Many beautiful varieties are seen in our gardens and parks, and even corn has a variety with striped leaves. They are easily reproduced, both by buds and by seeds, and they are the most ordinary of all varietal deviations. They may be expected to occur wild also. But no real variegated species, nor even good varieties with this attribute, occurs in nature. On the other hand occasional specimens with a single variegated leaf, or with some few of them, are actually met with, and if attention is once drawn to this question, perhaps a dozen or so instances might be brought together in a summer. But they never seem to be capable of further evolution, or of reproducing themselves sufficiently and of repeating their peculiarity in their progeny. They make their appearance, are seen during a season, and then disappear. Even this slight incompleteness of some spots on one or two leaves may be enough to be their doom.

It is a common belief that new varieties owe their origin to the direct action of external conditions and moreover it is often assumed that similar deviations must have similar causes, and that these causes may act repeatedly in the same species, or in allied, or even systematically distant, genera. No doubt in the end all things must have their causes, and the same causes will lead under the same circumstances to the same results. But we are not justified in deducing a direct relation between the external conditions and the internal changes of plants.

These relations may be of so remote a nature that they cannot as yet be guessed at. Therefore only direct experience may be our guide.

Summing up the result of our facts and discussions we may state that wild new elementary species and varieties are recorded to have appeared from time to time. Invariably this happened by sudden leaps and without intermediates. The mutants are constant when propagated by seed, and at once constitute a new race. In rare instances this may be of sufficient superiority to win a place for itself in nature, but more often it has qualities which have led to its introduction into gardens as an ornamental plant or into botanical gardens by reason of the interest afforded by their novelty, or by their anomaly.

Many more mutations may be supposed to be taking place all around us, but artificial sowings on a large scale, combined with a close examination of the seedlings and a keen appreciation of the slightest indications of deviation seem required to bring them to light.

Reproduced by permission of the publisher

LORANDE LOSS WOODRUFF 1879–1947

Foundations of Biology, New York: Macmillan, 1946.

GENETICS AND EVOLUTION

The modern approach to the critical analysis of significant variations was opened by the work of two botanists: de Vries and Johannsen. De Vries laid stress on the importance of discontinuous variations which he called *mutations*; while Johannsen made clear that in a homozygous germ complex, or *pure line*, selection is ineffective.

Some of the problems of selection will be clear from an example. Take, say, a quart of beans and sort them into groups according to the weight of each bean. Then put each group into a separate cylinder and arrange the cylinders in a series according to the weight of the enclosed beans. Now if we imagine a line connecting the tops of the bean piles in the cylinders, it takes the form of a normal *curve of probability*, or *variability curve*. A similar figure would be obtained by the statistical treatment of nearly all fluctuating characters among the members of any large group of organisms, or of the size of the grains in a handful of sand, or the deviations of spots from the bull's-eye in a shooting match. Therefore the variations with respect to a given character very closely approximate the expectation from the mathematical theory of probability, or chance, and the reasonable conclusion is that such finely-graded fluctuating variations are a resultant of a large number of factors, each of which contributes its slight and variable quota to the expression in a given individual.

The question is, what results are obtained by breeding from individuals which exhibit such a fluctuating variation to, let us say, a greater degree than that of the mean of a mixed *population*? With Galton's principle of filial regression in mind, one will naturally expect, and rightly, that the offspring usually will exhibit the character to a less degree than the parents but to a greater degree than the population. The top (mode) of the curve will have moved, so to speak, slightly in the direction of selection. Now, by continuing generation after generation to select as parents the extreme individuals, is it possible, with due allowance for some regression, to take one step after another indefinitely, or until the character in question is expressed to a degree which did not exist previously? The experience of practical breeders gives a partial answer, since the continual selection of the best animals for mating and the best plants for seed has been a profitable procedure. But it has long been known that after a certain amount of selection has been practised it may cease to be effective, and thenceforth serves chiefly to keep the character at the higher level attained.

The crux of the matter is in regard to exactly what the variations are. Both modifications and recombinations are usually included, and this mixture of non-heritable and heritable variations is what makes confusion. If we rule out recombinations by inbreeding or by self-fertilization of homozygous individuals, soon we establish *pure lines*. Then the variations are all modifications and selection is ineffectual with characters which are not inherited.

The importance of this point was discovered by Johannsen in careful experiments on the inheritance of characters in single pure lines of a brown variety of the common garden bean. For example, by keeping the progeny of each individual bean separate from that of all the rest, he was able to isolate a number of pure lines which differed in regard to the average weight of the beans. Thus selection resolved the bean population with which he began into its constituent "weight types," or lines, each of which exhibited a characteristic variability curve of its own with a mode departing more or less from that of the population. But when Johannsen selected *within* a pure line (ruled out recombinations) nothing at all resulted; he was unable to shift the mode because he was dealing with non-heritable characters. In other words, selection sorts out pre-existing pure lines (lines with homogeneous germinal constitution) from a population and then stops. Thereafter mutation must occur in the pure line for selection to be effective—but by the mutation the single pure line becomes two.

The pure line concept has served to clarify our ideas in regard to selection by focusing attention on the actual nature of the variations being dealt with. It discriminates sharply between modifications, which are a result of environmental influences, often recurrent in each generation and so *seemingly* inherited, and variations which are heritable because they are the result of changes in the germ plasm.

Reproduced by permission of the publisher

CHAPTER XI

PREFORMATION AND EPIGENESIS

THE ORGANIZATION OF THE GERM-CELLS

THE problem of the transmission of form in animal and plant species from generation to generation, and consequently of the evolution of the embryo, has puzzled observers throughout the ages. In the fourth century B.C. Aristotle had already contrived experiments relating to the development of the hen's egg. Until the discovery of the sex elements, however, such experiments could not be said to have been of a scientific nature, that is to say until 1667 when Leeuwenhoek discovered spermatozoa.

Spallanzini in 1754 and Prevost and Dumas some time later in 1824 showed the necessity of having spermatozoa in the fertilization process, since on filtering the sperm secretion of a frog prior to mixing it with the egg-cells, the latter failed to develop. Not until 1875 did Hermann Fol and Oscar Hertwig, working independently of each other, describe the penetration of sperm into the egg-cell and the formation of the zygote nucleus, the egg, by combination of nuclear chromatin from such sperm with that of the egg-cell nucleus.

The male and female sex cells must possess, therefore, organogenetic properties, since their progressive differentiation and increase gives place to the evolution of the embryo. The most direct theory could not help but be the assumption that in such cells there must be material or potential representations of every part of the organism. Hartsoeker in 1694, for instance, gave a sketch of a large-headed mannikin dwelling in the sperm chromatin, growing up to become the foetus and feeding on the egg substance at the beginning and then on sustenance directly supplied by the mother. Other writers of the eighteenth century held that a microscopic ready-formed being was to be found in the egg-cell whilst the sperm must be some sort of parasite absolutely superfluous as far as reproduction went.

The theory of preformation has had its partisans in all ages: Swammerdam, Bonnet, Haller, Malpighi and Buffon in earlier times and Baer, Remak, Kölliker, Haeckel and others in more recent centuries, and still closer to our own times Wilhelm Roux with his mosaic theory, not to mention many pure philosophers, particularly Malebranche and Leibnitz.

All writers who sought to explain the formation of the organs as the work of representative particles existing in germ-cells must be classed as preformationists. Such representative particles are presumably contained in the primary germ-cells and become liberated successively in the course of generations until one day or other the stock becomes effete, as mentioned by Bonnet for instance. Or they might be reproduced indefinitely by division in the same way as the cells containing them, as are also the genes and living molecules.

PHYSIOLOGY OF EMBRYONIC DEVELOPMENT

Opposed to such morphological ideas, of what we might call the anatomical type, are those of other biologists who consider development to be but one among many other functions. Forms, they say, appear through *epigenesis*, because in the embryo stage the circumstances necessary to the purpose arise and determinant causes come into play at the right moment, so that inevitably the organs naturally follow one after the other differentiating successively at each stage of development. In the seventeenth and eighteenth centuries particularly, disputes were constantly taking place between the preformationists and the epigeneticists. Wolff in 1759, at his inaugural lecture at Königsberg, maintained that embryogeny was merely a function and, referring particularly to the growth of domestic fowl, stated that it is by no means possible to agree that the germ will contain, even invisibly, all the elements of the mature creature which is distinguishable by the great amount of differentiation and tremendous complexity of its individual organs; embryogeny, he said, is something more than a mere growth from a number of preformed outlines. More than a century later, W. His, in 1874, published his work *Unsere Körperform und das physiologische Problem ihrer Entstehung* (The Shape of our Body and the Physiological Problem of Its Perfect Development). He agreed that organ-forming regions exist in the egg, but at any instant during the production of the embryo the first buds of the organs also start to make their appearance, each of them developing on its own account, whilst influencing and being influenced by the remainder. This is the first reference to ideas which are nowadays commonly accepted, commencing particularly with the work of W. Roux since 1883. Differentiation takes place—

(a) Through properties intrinsic in the germ-cells (auto-differentiation from the outlines), and (*b*) Through correlative differentiation, mutual influences between the various parts of the embryo, or dependent differentiation.

These two aspects correspond, as is evident, respectively to pre-formative and epigenetic mechanisms.

Driesch in 1895 gave preformation the credit of being of *prospective significance:* each part of the germ-cell or of its descendants will in normal conditions give rise to a particular portion of the organism. He also named as *prospective potential*, the epigenetic possibility that each morphological element, from the zygote or from the blastomeric aggregate, might generate organs different from those which they ordinarily produce.

It is easy to show how during embryogeny, as well as throughout the whole course of life, essential properties from the organized living matter help in morphogeny and function, and how, moreover, they influence conditions which follow one another as time passes on, outlining the vital course of a creature's existence.

ANISOTROPY IN THE EGG

Pflüger's error, for instance, in assuming that the egg was isotropic, that is to say, homogeneous in its formative determinates, has been demonstrated many a time and oft. It is incontestable that germ localizers, in His's sense, reside within the egg. There is no exception, even though the egg may present the most remarkable example, for the same thing occurs in all reproductive cells and is proportionate to the activity of reproduction.

The polarity of the egg, resulting from the polarity of the egg-cell, is easily verified. The ejection of the polar bodies, resulting from double division on ripening, takes place through a spot on the surface of the egg-cell thus determining the polarity of this latter. The point at which this takes place becomes the animal pole, whilst at the diametrically opposite end we get the vegetative antipole. In vertebrates, the pole is that part of the egg-cell adherent to the ovary; in invertebrates this becomes the antipole. From the blastomeres issuing from the polar hemisphere the ectoderm is formed, this being the source of the organs of animal existence, whilst those deriving from the antipolar half give us the endoderm or source of the vegetative organs. In the spot corresponding to the antipole, the blastopore develops, making way for the primitive digestive apparatus, the archenteron which ultimately becomes the gut cavity.

In a frog's egg-cell, polarity is indicated by pigmental arrangement;

the antipolar region is white, whilst the remainder of the egg-cell is dark, and transition from one zone to the other is observable through parallel strata located beneath its equator. There are other similar examples.

Anisotropy of the egg-cell becomes more evident at the instant of fertilization. The entry of the sperm causes profound alterations in the cytoplasm, and most certainly, in the nucleus, the structure thus obtained acting on the segmentation.

It is the rule that only one sperm can enter the egg-cell, whereupon the latter, after penetration has been achieved, acquires a genuine immunity and repels any subsequent sperm, in some cases even filming over adventitiously with an impermeable cover. Polyspermy is something out of the ordinary; it may possibly happen, for instance in damaged egg-cells subjected to the action of anaesthetics. Eggs very rich in yolk, especially telocytes, may become polyspermic, but only one single sperm, one single male pronucleus, will actually combine with the ovular nucleus. Other sperms will disappear and only under abnormal circumstances can polyspermic fertilization take place.

In the frog's egg-cell, a grey area appears immediately after fertilization below the equator and above the white antipolar cap. Its shade is halfway between the dark of the pole and the pale of the antipole, whilst its appearance is that of the moon's last quarter, or a "grey crescent" as Morgan termed it. Now, the symmetry of this crescent is determined by the point at which the sperm makes its entry, and its upper part corresponds to the longitude at which such entry took place. The plane of section of the sphere, i.e. of the egg, containing the pole, antipole and entry meridian, which is the axis of symmetry of the crescent on which the entry-point is located, will in seventy cases out of a hundred, according to Brachet, be the primary plane of cleavage for the egg when it divides into its two blastomeres, i.e. the sagittal plane; in 10 per cent of cases it will be the frontal plane, i.e. the perpendicular to the one last mentioned, and in the remaining 20 per cent it will take up an oblique position varying according to circumstances. In the first case, the crescent is shared in halves between the two first blastomeres; in the second, it is contained fully within one of the blastomeres, the other having no part of it; whilst in the third eventuality the division of the crescent will be unequal.

The primary plane determines the remainder. The successive planes of division are always perpendicular among themselves. If the division of the zygotes into two blastomeres takes place in the sagittal plane, the following division into four will occur on the frontal plane, the next into eight on the horizontal and so on. Roux's observations tell us the importance of the bearing of the planes in blastomeric division. If the cleavage plane in the zygote is sagittal, and one of the two blastomeres resulting from such segmentation is afterwards destroyed, only half an embryo is obtained; if the segmentation plane is frontal, the surviving blastomere will give either a whole embryo or an undifferentiated conglomeration of cells according to whether it comes from the part of the egg with the crescent or without it. Herlitzka's experiments in 1893 have supplied confirmatory data on this point, as did Spemann's ligature experiments in 1901, and Winterberger's more recent researches in 1929 on killing one of the blastomeres with X-rays. All this tells us how, before cleavage, the march of morphological events to develop in the genesis of the embryo has already been predetermined as if obeying some already well-laid master-plan.

What we have just described is not exclusive to the frog. Other batrachians and amphibians, anura and urodela, show analogous phenomena, even though the modification of the cytoplasm following on fertilization may not be so evident to the sight, since the grey crescent does not show itself in these species.

THE SIGNIFICANCE OF BLASTOMERIC DIVISION

Boveri found, on the other hand, that in the sea-urchin, after fertilization, a pigmented ring made its appearance above the equator of the egg, and had a definite connection with the division into blastomeres. At the "eight" stage, that is the third division, the ring appears in four antipolar blastomeres and at the bottom of the four polar ones. At the next or "sixteen" stage, when the blastomeres are no longer equal in size, the eight upper polar ones of medium size, or mesomeres, have the lower portion pigmented and the main part of the ring surrounds the four blastomeres situated beneath them, i.e. the larger blastomeres or macromeres; whilst the four antipolar blastomeres which are the smallest, or micromeres, are quite colourless. In the sea-urchin, the invagination which will form the gastrula commences with a flattening of the antipole, and then the cells deriving from the micromeres drop into the blastocele or segmentation cavity where the mesenchyme or primary mesoblast will develop. The portion which is invaginated and forms the source of the archenteron comes from pigmented macromeres. In this fourth cleavage it is still possible to tell the origin of the cells forming the essential parts of the gastrula and the primitive streaks, thanks to the localization of pigment. It is found, also, that blastomeres continuing to multiply, like the organs which will later develop from them, are already roughly outlined in the egg, even before any cleavage whatsoever.

In certain Ascidians, *Styela partita*, Conklin was able in 1905 to describe similar occurrences. Insemination of the ovule similarly provoked alterations in the morphogenic structure thereof. In a particular portion of the surface a yellowish crescent rich in lipoids develops, corresponding to the region which later becomes the rear region of the embryo, and it is easy to relate the cells deriving from this region, as far as to the gastrula and still farther even, since they are distinguishable by their pigmentation.

Likewise in *Amphioxus*, Conklin showed in 1932 that already in the first stages of cleavage, up to the fourth (with sixteen subdivisions), it is possible to define the cells from which the different parts of the gastrula proceed. Bilateral symmetry is determined in the first cleavage, and the dorsal and ventral faces in the second. As we have seen, morphogeny develops with precision, starting at the first instant of the blastomeric division. Artificial marks, similar to these natural markings, which are found in certain species, can be imprinted artificially by experiment, and this will help us likewise to follow the route and destination of strings of cells during embryogeny. This was the idea W. Vogt had in 1925, when he suggested the use of stains. Various dyes, which of course must be indelible, can be used, for instance Nile blue, neutral red, Bismarck brown, bright blue, etc. (Lehmann, Weissenberg and Winterberger). Ansell and Winterberger used electrolytic markings. Such staining allows us to study the fate of various portions of the egg, of different blastomeres and so on right up to a fairly advanced stage in development.

Vogt made use of various kinds of batrachians; von Ubisch and Hörstadius worked with sea-urchins' eggs, and Teissier with hydra. It has been possible to establish the origin of vestiges which do not show themselves until later, thus proving that each species· has a definitely fixed characteristic scheme of development. It has been proved, moreover, that relative movements of the cells during increase, spiral cleavage and cell migrations do take place, in other words that there are such things as morphogenetic motions which are quite contrary to Oscar Hertwig's original idea.

PREDETERMINATION IN THE EGG

It is quite evident that there is predetermination in the egg, arising from the ovule and from the sperm, though it is neither inviolable nor decisive. Countless researches have proved that if during the first instants of embryo formation a blastomere or group of blastomeres be destroyed, there are sufficient others to take their place in the general development.

Investigations in this direction have been numerous, and contradictory results have at times been obtained during their course; W. Roux in 1886 used a red-hot needle to destroy one of the first-stage blastomeres from a frog, and got half-embryos both right and left. These anomalous embryos, nevertheless, went on to full development later with more or less conformity to standard according to circumstances. Chabry in 1887 likewise killed some of the initial blastomeres by pricking, working with Ascidians, and as in Roux's case, the result was a semi-embryo. Conklin in 1901, a considerable time after this, was able to corroborate the same thing experimenting with the ascidian *Styella*.

On the other hand, Driesch (1892–1900) reached important conclusions to directly the opposite effect. Isolating the two primary blastomeres of a sea-urchin, he obtained two complete larvae, though they were much smaller than normal. He observed, however, that as the development of the embryo proceeded, the blastomeres went on gradually losing their total power, until this vanished entirely in the fourth or fifth cleavage. From then on the blastomeres ceased to be equivalent among themselves and to possess the same faculties. This has been confirmed by various writers—Wilson, Zoja, Yatsu, Zeleny, Herlitzka, Conklin, Morgan and others—using various procedures for isolation, such as destruction of one of the blastomeres, ligature (Herlitzka, Spemann), dissection under the stereoscopic microscope (Driesch, Hörstadius), pressure with a thin glass rod (Mangold), gentle stirring in water (Driesch, Boveri, Wilson, Conklin), contact with calcium-free brine or sea water (Herbst), and so forth.

Merogony, cleavage of the blastomeres, gives different results not only according to the moment of development at which it takes place but also according to species. Thus, in Ctenophora, Driesch and Morgan and Fischel and Yatsu noticed that blastomeres isolated in the first, second and third cleavages gave respectively half, quarter and one-eighth of a normal larva. Crampton in 1896 mentioned that the same happens in certain gastropod molluscs, Wilson found the same in Scaphopoda (elephant's tusk shells), and Conklin in another mollusc the Crepidula. Chabry demonstrates similar results in Annelids and Ascidians.

MOSAIC EGGS AND REGULATION EGGS

There are, therefore, some eggs which behave as if the functional and morphogenic traces were working in a definite way, whilst others, on the contrary, have blastomeres giving place to organs which they would not form under normal conditions. There are some eggs in whose development self-differentiation prevails, and others wherein correlative differentiation, in Roux's sense, is predominant. The former are called eggs of *real potential* in mosaic, and the latter *regulation* eggs of *full power*.

Blastomeres proceeding from the first cleavage of an egg in mosaic give birth to incomplete embryos, such as we have seen occurring in the case of certain Ascidia and Annelidae (Chabry), in frogs (W. Roux), in Ctenophora (Driesch and Morgan). In regulation eggs, each blastomere produces a complete embryo, as in the sea-urchin (Driesch), in Amphibia (Herlitzka), and in Amphyoxus (Wilson). This shows that a local influence such as position in the germ of those organo-genetic traces of which W. His spoke, may be possible and even coincide with epigenetic factors which make themselves evident during development. These are two aspects of embryogeny which are not evidenced with equal effectiveness in each species—since both mosaic and regulation eggs exist—nor at every instant during ontogeny. Generally total power prevails at the start of development and weakens progressively as the contours of the organs become plainer. The egg is not a rigid prototype of the embryo since parts are also determined by interaction. This in no way means that the moments of morpho-genetic evolution are not genetically imposed right from the germ-cells. It does not seem difficult nowadays to reconcile these two apparently opposed points of view, which divided embryologists so long ago into two camps.

Recent advances in experimental embryology have, however, led to results of the greatest importance. We must specially mention researches based on the union of blastomeres proceeding from different zygotes, either of the same species or each from different species. Hörstadius in 1928, working on blastomeres, fragmented from sea-urchin embryos, obtained pluteus larvae of different form according to graft. He was able to "animalize" or "vegetalize" at will any particular part of the embryo, by adding suitable complementary portions thereto. These results were confirmed by fresh researches in 1936 using *Nemertea* (ribbon-worm) eggs. Von Ubisch (1925-9) completed similar experiments but with the main object of obtaining *chimeras* or graft hybrids by uniting two distinct species in various combinations. In some cases the grafted elements are eliminated; in others they develop in more or less incomplete and isolated groups within a single embryo, derived from each one of the original blastomeric groups; in yet others, however, regulation phenomena are in fact observed tending to accommodate the whole to make up an individualized system which will indeed be a true crossbreed.

On a like foundation with the foregoing are based the researches of Spemann and his school, who implanted parts of one embryo in another host embryo. This method consists in alternating fragments *a* of embryo *A* with portions *b* of the second embryo *B*; thus *a* is inset in *B* and *b* in *A*, changing their anatomic disposition. It will be seen whether the grafts *a* and *b* in their new position develop in the same way as they might have done in their original location, i.e. in accordance with their origin and by virtue of their inherent properties "preformational self-differentiation" or if they develop under the influence of their new location "correlative differentiation."

EPIGENESIS AND PREFORMATION VARYING WITH TIME

These researches have led to a confirmation that in the initial moments of development, the total power of the blastomeres is greater than the actual potentiality, in other words that epigenesis is at that stage predominant. The fate of each one of the parts of the embryo is not at that moment irrevocably determined and extensive regulation generally can set in. As embryogeny proceeds this capacity for regulation diminishes more and more.

The work of Spemann and his co-workers was carried out on Urodela and various species of tritons. The original test consisted in changing grafts between two young gastrulas of like age, namely fragments of epidermis, one of which, *a*, was taken from the region in front of the blastopore which later becomes the neural plate, whilst the other, *b*, came from the latero-ventral region where skin develops naturally without any special physiological significance. Graft *b* will therefore give birth to nerve elements, whilst *a* on the other hand will give skin cells. This shows that the fates of the *a* and *b* fragments were not definitely determined at source, but that morphogenic fixation springs from the spot where the grafts begin to develop locally.

Later, Spemann carried out further, more convincing tests, still using two different species, namely *Triton taeniatus*, whose eggs are highly pigmented, and *Triton cristatus* with light-coloured eggs, the eggs giving similarly coloured embryos, dark and light respectively. In this way plainly visible chimeras are obtained. The pigmented graft stands out as a dark stain on the light embryo, and conversely. Depth of section is also easily defined by the presence or absence of pigmentary granulations in the cells of the grafted tissues.

This technique has enabled us to make some very interesting deductions. At the start of gastrulation, the antipolar hemisphere can be replaced, through implantation, by the polar, provided that the section be made above the zone corresponding to the crescent. The meridians of the cap and the remainder of the blastula may be allowed to coincide, or the cap may be given a twist of one or two right angles at the time of positioning for implantation. A normal embryo is obtained; the neural plate in this case develops before the blastopore, just as if there had been neither substitution nor rotation, and this occurs at the expense of cells which would normally have generated skin and not nerve elements. If the horizontal section be made lower down, so as to affect the crescent zone, and a rotation of 180° be given, a double embryo will be obtained, one of whose elements differentiates in its normal position, i.e. in the dorsal surface of the egg, and the other at 180° in the ventral surface. Under such conditions, working only with comparatively fresh gastrulas, we get proof that the fate of the polar hemisphere is not determinate up to this point.

On the other hand, in the neurula stage, the outlines have already been definitely differentiated. Grafts therefore do not become differentiated according to location, but in accordance with their source. The moment of change from epigenesis to determinancy occurs in this species between the end of gastrulation and the formation of the neural plate.

The existence of germ localization, anisotropy, is consistent with the extremely extensive regulational capacity of certain eggs. Spemann's own work proved that even in the first cleavage, there is an inductor centre conditioning the formation of the blastopore. This blastopore later becomes a gastrular localization endowed with high inductional powers. Isolation of blastomeres in the second and third cleavages causes only those to gastrulate which contain the necessary substance for forming a blastopore.

INDUCTORS AND ORGANIZERS

Grafting, as in Hilde Mangold's experiments, a fragment from the anterior lip of the blastopore of one species of Triton, say the pigmented type, to the latero-ventral region of the clear embryo from another species, will determine in the host the development of a secondary embryo characterized by axial organs, spinal chord, nervous system, etc., and incite the "neural organization of tissues" which, without its influence, would have given rise to other organs more usual to that region. The only active factor is the anterior lip of the blastopore, which not only brings about differentiation of suitable tissues but also that of the tissues of the host. The graft may be said to induce the formation of organs in the host.

On the other hand, if we implant other tissues not from the anterior blastopore lip, they are induced by the neighbouring host tissues and contribute to formation of the organs which should be differentiated in the locality under consideration.

The effects of the graft of an organizer are definite evidence of epigenic influences in the course of development, the organizer seems to be some one or several chemical substances and shows no zoological specificity. Inductive action can be transmitted by contact.

There are other inductions of a different category; for instance, primary inductions which are the most intense and come from the blastopore lip in the first instance, secondary and remoter ones which are more and more localized and subordinate. The case of development of the crystalline lens of the eye, studied by Spemann in 1901, Lewis in 1904, King in 1905, and still later by Filatov in 1925, shows yet again that organs become differentiated thanks to a double hereditary influence, preformational and epigenically inductive, a common twofold instrument called a "double safety device" by Spemann.

We could mention other experimental results to corroborate the foregoing. Functional and morphological characters are certainly already foreshadowed in the egg by the genetic factors contained therein, without which inheritance would become impossible. This implies determinancy in the ovule, and thus there is no doubt that organism results from properties in the egg itself. The parts, however, which go on to production are determined by mutual action and this allows regulations which will be more or less far-reaching according to species.

Influences are exercised, as we previously saw, by material agents—inductors, organizers, hormones, metabolites, and so on—and perhaps also by as yet undetermined dynamic factors acting nevertheless in fields of force, in a well-defined spatial framework, and evidenced by dynamic gradients as mentioned by Child in 1921, the importance of which latter has been dwelt upon by Huxley and de Beer (1934).

It can be deduced from all this, that in embryonic development, the same processes of correlation and organic unification are at work as in other biological and physiological phenomena in general: elementary properties of the cell in the first place and chemical and energetic integrated mechanisms in the second. A perfect adjustment between the elementary properties and the functional processes of differentiated parts to form an integral whole or organism makes possible the development of life in its most diverse aspects, the most remarkable of which at present seem to be instinct on the one hand and on the other embryological or reproductive morphogeny and the auto-regeneration of morphological characters. Constantly, even in the most distant reaches of biology, we find properties of cell structure and correlation between cells all in synergetic activity. In development, the former is preformation and the latter epigenesis.

It is therefore unjustifiable in these days to carry on ancient feuds. Recent advances have shown us that in ontogenic evolution, preformation and epigenesis exist side by side. Furthermore, they have been shown to be one and the same. Everything has its own moment in life. In the formation of a new body, everything lies naturally implicit in the germ-cell. The offspring unfailingly follows its set course, existence, and this course will be one of many biological properties distinguishing the individual and the species. Preformation does not only assume certain forms, but assumes that certain instants also are of importance. When the time is due, a particular phenomenon will make its appearance, and this may be the creation of a bodily part just as much as of a function. Also, among so many functions, we have the production of one particular substance or the liberation of a force, and therefore, a relation among the cells and their mutual influence; in short—epigenesis. Epigenesis is therefore just one more piece of evidence, amongst so many, of genetic preformation, which in its turn is not mere anatomy.

*　　*　　*　　*　　*

JULIAN SORELL HUXLEY *b.* 1887 *and*
GEORGE RYLANDS DE BEER *b.* 1899

The Elements of Experimental Embryology, Cambridge: University
Press, 1934.

It is now possible to give a brief summary of the chief points which have
emerged from our study of development, during which attention was
focused on differentiation and its origin as the central problem.

In the first place, animal development is truly epigenetic, in that it
involves a real creation of complex organization. It is also predetermined, but only in the sense that an egg cannot give rise to an organism of a species different from its parent. The development of each
individual is unique. It is the result of the interaction of a specific
hereditary constitution with its environment. Alterations in either of
these will produce alterations in the end result.

Determination is progressive. In the earliest stages, the egg acquires
a unitary organization of the gradient-field type in which quantitative
differentials of one or more kinds extend across the substance of the egg
in one or more directions. The constitution of the egg predetermines
it to be able to produce a gradient-field of a particular type;
however, the localization of the gradients is not predetermined, but is
brought about by agencies external to the egg. The respective roles of
internal predetermination and external epigenetic determination
are clearly seen in regard to the bilateral symmetry of the egg. The
amphibian egg is predetermined to be able to give rise to a gradient-
field system of bilateral type through the establishment of the grey
crescent at a particular latitude of one meridian. The particular
meridian is not predetermined, but is normally decided by the point of
sperm-entry; the precise latitude is determined as a result of the
primary axial gradient of the egg, impressed upon it by factors
in the ovary. On the other hand, the egg of a radially symmetrical
animal like a Hydroid is incapable of developing bilateral symmetry;
the predetermined capacity to react to stimuli localized in one meridian
is not given in its constitution.

The agencies which determine the position of the various axes
involved in the gradient-field system may be of very various nature;
they may be factors in the maternal environment (ovarian conditions),
biological factors (point of sperm-entry), or external physical factors

(as in the determination of the polarity of the egg of *Fucus*). In any case they are external to the egg. They may also operate at very different times relatively to fertilization.

A number of chemical processes are set going by fertilization. These will proceed differently in the quantitatively different environments provided in different parts of the gradient-field system, until qualitative differences are set up. In most cases, these differences are at first not visible, and are presumably of chemical nature; this step in differentiation is therefore spoken of as chemo-differentiation. These chemical differences appear at first to be reversible (e.g. labile determination of the presumptive neural tube region in the Urodele before gastrulation) but after a certain point to become irreversible. From this moment onwards, the organism consists of a mosaic of chemo-differentiated regions, each determined to give rise only to one or a limited number of kinds of structure. These are what we have called partial fields.

The attainment of the mosaic stage often takes place under the influence of a dominant region or organizer. This may determine the extent and form of the whole gradient-field within which chemo-differentiation occurs, as in Planarian regeneration, or may interact with a previously established gradient-field orientation in another direction, as in amphibian organizer grafts.

The organizer may exert its effects at a distance, as does the regenerated head on a cut piece of a Planarian, or may supplement such distance effects by more powerful contact effects, as happens when the amphibian organizer comes to underlie a certain portion of the animal hemisphere, and at once determines it irrevocably as a nervous system.

Modifications of the gradients by external agencies will entail alterations in the structures produced. These alterations may consist in changed proportions, or in the total absence of certain regions (temperature-gradient experiments with frogs' eggs, cyclopia in fish, modification of regeneration in Planarians). Here again, there is a predetermined capacity to produce a certain type of structure in certain conditions; but the precise localization of the structures produced depends upon the form of the gradients in the field-system.

Once the mosaic stage has set in, further differentiation may be brought about by the influence of one point on its neighbours. The classical example of this is the induction of a lens from epidermis by the optic cup.

During the period when the organization of the developing animal consists of a single field-system, far-reaching regulation is possible; after irreversible chemo-differentiation has occurred, it is not. The precise time at which irreversible chemo-differentiation sets in varies markedly in different groups. In Amphibia it occurs during gastrulation; in Ascidians at fertilization.

After the establishment of a mosaic of partial fields, it does not follow that all the cells of any given partial field necessarily give rise to the organ characteristic of the field. Thus, more cells are capable of giving rise to the amphibian fore-limb than do in fact give rise to it in normal development. Further, the boundaries of the partial fields overlap: a given group of cells in the limb-rudiment of the chick may contribute to the formation of either a thigh or a shank, according as to whether it is allowed to remain attached to or is isolated from one partial field or the other. Gradients may exist in such fields: the capacity of cells within the fore-limb field to give rise to a limb decreases with their distance from a subcentral portion of the field: the same is true for many other organ-fields.

Up to a certain time, regulation is still possible within each of the partial fields; but as development proceeds, each of these becomes split up into progressively smaller fields, each with its own determined fate: for instance, the fields for leg, shank, and foot, within the originally single hind-limb field.

Each area in the mosaic passes from the state of invisible *chemo-differentiation* by the process of *histo-differentiation* to full visible differentiation, and so reaches the functional stage. After the organism as a whole has reached the functional stage, many new morphogenetic agencies come into play. The organism also, through acquiring the power of regeneration, re-acquires much of the regulative capacity which it lost in its passage through the mosaic stage.

The type of organization characteristic of one stage appears to persist, in whole or in part, throughout subsequent stages. Thus, the main gradient-system of the embryo permeates the partial fields of the limb, neural folds, ears, gills and heart, and determines their axis; and the growth of the lateral line along a particular level of the flank can best be interpreted in terms of a persistent total gradient-field.

Again, a total field-system certainly exists in adult Planarians and appears to reveal its presence in late stages of other groups through the presence of growth-gradients permeating the whole organism.

The persistence into adult life of the partial field-systems of the mosaic stage is shown by the phenomena of regeneration, by the existence of localized growth-gradients within single areas, and notably by phenomena such as those found in newts, where, for instance, indifferent regeneration-buds produced by an amputated limb will produce legs when grafted into a certain area round the leg, while if grafted near the base of the tail they will produce tails.

With this, of course, only a start has been made with the scientific analysis of development. It remains for the future to discover such fundamentals as the physiological bases of the field-systems, and the elaborate physico-chemical processes which must be operative at the time when the quantitative differences of the early gradient-field system are converted into the qualitative differences of the chemo-differentiated mosaic stage.

It is, however, already a good deal to have arrived at this first outline of development on the biological level. To have established the fact that organizations of quite different type succeed one another during development is important. The recognition of the gradient-field system, with its purely quantitative differentials, as the basis of early organization, is a great step forward, since it provides an adequate formal explanation of many phenomena of regulation which have been considered by various authors, notably by Driesch, as affording proof of vitalistic theories of development.

Further, the epigenetic analysis of development is pointing the way to a large extension of the field of heredity, in the shape of physiological genetics. It is only through a study of development that it will be possible to understand what the term "genetic characters" really stands for—in other words, what are the basic processes involved in the action of a particular Mendelian gene.

Experimental embryology as a separate branch of science was initiated by Roux; in its next phase, in which Driesch, Boveri, Wilson, Herbst, Morgan, Brachet and Jenkinson are outstanding names, a large body of facts was amassed, and the experimental proof of epigenesis provided; in the third phase, Spemann and Harrison are the outstanding figures within the sub-science, while the theories of Child have not only linked the facts of regeneration with those of embryonic differentiation, but have provided a scientific basis to early development, thus filling a large gap in the theoretical aspect of the subject.

Meanwhile, experimental embryology has been making fruitful contacts with physiology, notably in the field of hormone action, with genetics, and with growth studies.

The fourth stage is now beginning, in which this framework of general principle will be filled in through intensive research, and the whole science deepened by a search for the physico-chemical bases of the empirical biological principles which have been discovered in its earlier stages.

Reproduced by permission of the publisher

MAURICE CAULLERY *b.* 1868

Les Progrès récents de l'Embryologie expérimentale (Recent Advances in Experimental Embryology), Paris: Flammarion, 1939.

The question of differentiation in the embryo lies in the field of chemistry and perhaps more particularly in the province of physico-chemistry when we come to the study of general problems relating to nutrition and growth. Such a basis will also furnish a useful insight by means of the investigation of reproduction in the lower single-celled organisms, in which, though it cannot be said there are actually any embryological problems, we do discover cell reactions of great value. At the present time, for instance, several investigators are attempting the culture of single-celled species, particularly of flagellata, in accurately determined conditions of the nutritional environment. By such means we shall be able to decide what may be the most satisfactory substances for fostering the multiplication of such organisms, by "factors of growth" as they were named by Lwoff in 1932. Furthermore, data obtained from the examination of tissue cultures are also of major importance.

Experimental researches along independent lines are likely to converge and coincide, thus setting in relief phenomena of a general order which are of fundamental importance in the existence of cells and therefore in embryology. Here again we may observe the mutual dependence of essential problems which when they make their appearance are at first glance quite unrelated, though upon going into them more deeply, we find that they are distinct manifestations of the same essential properties.

Morphological epigenesis, of which the extent of its influence has been shown by experimental embryology to have direct natural consequences in *physiological* epigenesis, has been rightly stressed by Wintrebert during recent years. The one is complementary to the other. Development leads to a properly constituted functional organism. The embryonic period has generally been held to be antithetic to the rest of the individual's existence. Functional existence commences indeed in the egg. The functions of the embryo correspond to morphology in its various stages; there is an exact relation between form and function at any given instant, and things cannot be otherwise. Each stage in embryogeny conditions the succeeding one, and that is the reason why temporary organs may appear quite different from those occurring in the adult with its particular activities. This is remarkably evident in those species which pass through larval stages and metamorphosis. I have tried to stress the idea that physiological epigenesis is demonstrated by experimental embryology. It is precisely because of its functional reactions that the embryo continues to evolve during its earlier stages of development as well as in the more advanced periods of its existence.

Physiological research, on the other hand, is not concerned with form, except as a sideline to function, and in the primary stages where the functions are not so developed, the general factors governing cell life become manifest.

This lends interest to the problems of experimental embryology, in which we can analyse the most interesting factors in the processes of vital activity, those of the widest effect, namely the cell processes in their most profound aspects. In every biological manifestation, the intrinsic properties of the cells take first place. After these come the phenomena resulting from relationships among the cells, and this is confirmed yet again in the development of the individual organism.

Reproduced by permission of the publisher

CONRAD HAL WADDINGTON *b.* 1905

Organizers and Genes, Cambridge: University Press, 1940.

A complementarity to the ideas of physics and chemistry is often alleged not only of life, but of the notion of biological organization. Indeed, in recent years, there has been a tendency either to regard organization as one of the irreducible fundamental bases of all biology, or to invoke it, as though it were a well-defined concept, to fill up any awkward gaps in a theoretical structure. The latter use is perhaps the more reasonable; but it is obviously dangerous without a clear idea of exactly what is meant by the term, and when an attempt is made to reach such clarity, I think it will be found that the scope of the notion, as an explanatory principle, is not so great as has sometimes been suggested. On the other hand, it does provide the key to an extremely valuable method of thought. But before it can be accepted as a guide, it requires more precise formulation than it has yet received, at least in scientific contexts.

In the first place, it is clear that the fundamental notion is organization, which is capable of quantitative variation, rather than organism, which is not. The degree of organization of an entity is usually considered to depend on the degree to which the parts of the entity are dependent on the whole. Now it is immediately clear, but is rarely pointed out, that the degree of dependence will be different in respect of different properties of the parts. For instance, if we consider a certain mechanism, such as a motor-car engine, the parts such as the cylinders, pistons, etc., are not in any way existentially dependent on the whole; they can exist perfectly well in isolation from it. They are, however, highly dependent on it in a certain context, namely with reference to the functioning of an internal combustion engine. When we speak of the dependence of the parts on the whole we must always have in mind some particular context; thus the parts of an entity can be said to be dependent on the whole, in a particular context, if, in order to express the properties of the parts in that context, some reference to the whole is necessary. For example, the eye may be considered either with reference to the function of vision, or as a developing entity; in the first context, the relations of the retina and lens are those of two bodies which have to be adjusted to one another in order to focus light, etc., while in the second their relations are those of an inductive reaction and the mutual adjustments of growth rates. There is obviously no reason why the internal co-ordination of the eye, or, to put it in the way previously used, the dependence of the parts on the whole, should be the same in the two cases. The eye may, and at certain periods of its existence undoubtedly does, have very different degrees of organization in these two contexts.

The purpose for which the concept of organization is usually invoked is to form part of the theoretical system for dealing with phenomena which seem to involve the subjection of otherwise self-sufficient parts to some overriding whole. For instance, one might be tempted to advance "the organization of the eye-forming region" as an explanation of the formation of a normal eye after some experimental disturbance of its rudiment. Instead of the individual cells behaving independently, they are subordinated to a general influence which causes them, even after disarrangement, to form a single normal organ. This is often expressed by saying that the tissue can no longer adequately be regarded as a mere mass of cells, but has attained a higher level of organization in which the relevance of the whole organ to the constituent cells can no longer be disregarded.

In some ways this type of expression suffers from the same difficulties as we noted in discussing fields. Where does the organization come from? And what is meant by a higher level of organization? The second of these questions is easier to answer. The statement made earlier that organization must be defined with reference to some context provides the clue. A new level of organization is in fact nothing more than a new relevant context. When it is said that an organ rudiment has a higher level of organization, as a developing entity, than a mere mass of cells, what is meant is that some organ is relevant to the former, while the latter has nothing to do with an organ, either because it is not yet competent or because it has passed the regulative stage and reached a point at which its development is completely mosaic, each fragment differentiating on its own without any reference to the whole.

The question of whence a new level of organization is derived is more delicate. In discussing the analogous question about fields, we stated that the field must be regarded as a product of the interaction of its parts. If we applied this directly to the problem of organization we might seem to arrive at the conclusion that all levels of organization are implicit in the levels beneath them, which constitute their parts; and that would lead to denial of any true arising of new contexts, and thus to a denial of levels of organization. In fact, we should find ourselves in a vicious circle. We do, however, seem to be confronted with the fact, which is very difficult to deny, of new contexts and new levels of organization. It does seem that we are thinking in terms which are in some ways at different levels when we think of genes and their immediate products, and of the mirror symmetries of reduplicated legs. Or, to take a more extreme case outside the present field, psychology is hardly relevant in connection with the chemistry of sugars.

We must then accept the existence of different levels of organization as a fact of nature. On the other hand, we cannot easily suppose that the arising of a new level involves the appearance out of the blue of completely new properties of the elements of which the organized unit is composed. But the way out of the dilemma is clear. When elements of a certain degree of complexity become organized into an entity belonging to a higher level of organization, we must suppose that the coherence of the higher level depends on properties which the isolated elements indeed possessed but which could not be exhibited until the elements entered into certain relations with one another. For instance, we have suggested that it may be possible to explain the organization of regions of embryonic tissue into organ rudiments by supposing that they contain some orientated protein microstructure similar to that of a liquid crystal. Such a microstructure is on a different level of organization to that of ordinary molecules in solutions. It depends on the mutual attractions and repulsions of fibre-like particles, which cannot be exhibited unless the molecules first form elongated fibres which then come together in large numbers. But the existence of the fibre level of organization is not accounted for in terms of the elementary molecules plus some entity of a higher level, such as an overriding field. Instead we account for the formation of the fibre field by enlarging our ideas about the elementary molecules to include the fact that they can polymerize into fibres, which will then attract and repel one another in certain ways. That is to say, a new level of organization cannot be accounted for in terms of the properties of its elementary units as they behave in isolation but is accounted for if we add to these certain other properties which the units only exhibit when in combination with one another.

According to this view, one cannot explain any peculiar behaviour by postulating the existence of a new level of organization. The advantages of the concept are not explanatory, but simply that it provides a terminology in which it is easy to admit the recognition that phenomena do not present themselves as being all of the same kind, and this is certainly an important advantage. In its absence we are practically forced to argue that the phenomena of sociology and chemistry are not significantly different, a point of view which cannot easily be made very plausible. The admission that different levels of organization exist frees us from such preconceptions. We feel no conviction that, for instance, the behaviour of a mass of tissue must be explicable in terms of the properties of its isolated cells. Instead, we hope that investigation of the tissue will reveal new data about the mutual interactions of the cells when aggregated in a mass. Our aim is not merely to explain the complex by the simple, but also to discover more about the simple by studying the complex.

EDMUND BEECHER WILSON 1856–1939

The Cell in Development and Heredity, New York: Macmillan, 1925.

CHROMOSOMES AND DETERMINATION

In what sense can the chromosomes be considered as agents of determination? By many writers they have been treated as the actual and
even as the exclusive "bearers of heredity"; numerous citations from
the literature of the subject might be offered to show how often they
have been treated as central, governing factors of heredity and development, to which all else is subsidiary. The most complete example of
this conception, perhaps, is embodied in the theory of the germ-plasm
as developed by Weismann; but in one form or another it has persisted
almost to the present day. Many writers, while avoiding this particular
usage, have referred to the chromosomes, or their components as
"determiners" of corresponding characters; but this term, too, is
becoming obsolete save as a convenient descriptive device. The whole
tendency of modern investigation has been towards a different and
more rational conception which recognizes the fact that the egg is a
reaction-system and that (to cite an earlier statement) "the whole
germinal complex is directly involved in the production of every
character." Genetic research is constantly bringing to light new cases
of the co-operation of several or many factors in the production of
single characters (e.g. in that of sex); and it is possible that all the
chromosomes, or even all of the units which they contain, may be
concerned in the production of every character. Beyond this it is
evident that every character is produced during development by an
activity in which the cytoplasm, and what we call the "organism as a
whole" plays a most important part. When, for example, we speak
of the X-chromosome as a "determiner" of sex, we mean only that
it is a differential factor or modifier the relative quantity of which in
relation to the autosomal material conditions a particular reaction by
the developing germ. The value of the chromosome-theory of
heredity does not lie in our identification of this or that "determiner"
or "bearer of heredity," but in its practical importance as a means of
experimental analysis. In this respect, in the writer's opinion, the theory
has the same kind of value as the molecular and atomic constructions of
physico-chemical science; and the "mystical" and "unscientific"
character ascribed to it by some writers is purely imaginary.

EDMUND BEECHER WILSON 1856–1939

The Cell in Development and Inheritance, New York: Macmillan, 1896.

THE UNKNOWN FACTOR IN DEVELOPMENT

We are now arrived at the furthest outposts of cell-research and here we find ourselves confronted with the same unsolved problems before which the investigators halted. The first question raises once more the old puzzle of preformation or epigenesis. The pangen hypothesis of de Vries and Weismann recognizes the fact that development is epigenetic in its external features; but like Darwin's hypothesis of pangenesis, it is at bottom a theory of preformation, and Weismann expresses the conviction that an epigenetic development is an impossibility. He thus explicitly adopts the view, long since suggested by Huxley, that "the process which in its superficial aspect is epigenesis appears in essence to be evolution in the modified sense adopted in Bonnet's later writings; and development is merely the expansion of a potential organism or 'original preformation' according to fixed laws." Hertwig (1892), while accepting the pangen hypothesis, endeavours to take a middle ground between preformation and epigenesis, by assuming that the pangens (idioblasts) represent only *cell-characters*, the traits of the multicellular body arising epigenetically by permutations and combinations of these characters. This conception certainly tends to simplify our ideas of development in its outward features, but it does not explain why cells of different characters should be combined in a definite manner, and hence does not reach the ultimate problem of inheritance.

What lies beyond our reach at present, as Driesch has very ably urged, is to explain the orderly rhythm of development—the co-ordinating power that guides development to its predestined end. We are logically compelled to refer this power to the inherent organization of the germ, but we neither know nor can we even conceive what this organization is. The theory of Roux and Weismann demands for the orderly distribution of the elements of the germ-plasm a prearranged system of forces of absolutely inconceivable complexity. Hertwig's and de Vries's theory, though apparently simpler, makes no less a demand; for how are we to conceive the power which guides the countless hosts of migrating pangens throughout all the long and complex events of development? The same difficulty confronts us under any theory we can frame. If with Herbert Spencer we assume the germ-plasm to be an aggregation of like units, molecular or supra-molecular, endowed with predetermined polarities which lead to their grouping in specific forms, we but throw the problem one stage further back, and, as Weismann himself has pointed out, substitute for one difficulty another of exactly the same kind.

The truth is that an explanation of development is at present beyond our reach. The controversy between preformation and epigenesis has now arrived at a stage where it has little meaning apart from the general problem of physical causality. What we know is that a specific kind of living substance, derived from the parent, tends to run through a specific cycle of changes during which it transforms itself into a body like that of which it formed a part; and we are able to study with greater or less precision the mechanism by which that transformation is effected and the conditions under which it takes place. But despite all our theories we no more know how the properties of the idioplasm involve the properties of the adult body than we know how the properties of hydrogen and oxygen involve those of water. So long as the chemist and physicist are unable to solve so simple a problem of physical causality as this, the embryologist may well be content to reserve his judgment on a problem a hundredfold more complex.

The second question, regarding the historical origin of the idioplasm, brings us to the side of the evolutionists. The idioplasm of every species has been derived, as we must believe, by the modification of a pre-existing idioplasm through variation, and the survival of the fittest. Whether these variations first arise in the idioplasm of the germ-cells, as Weismann maintains, or whether they may arise in the body-cells and then be reflected back upon the idioplasm, is a question on which, as far as I can see, the study of the cell has not thus far thrown a ray of light. Whatever position we take on this question, the same difficulty is encountered; namely, the origin of that co-ordinated *fitness*, that power of active adjustment between internal and external relations, which, as so many eminent biological thinkers have insisted, over-shadows every manifestation of life. The nature and origin of this power is the fundamental problem of biology. When, after removing the lens of the eye in the larval salamander, we see it restored in perfect and typical form by regeneration from the posterior layer of the iris, we behold an adaptive response to changed conditions of which the organism can have had no antecedent experience either ontogenetic or phylogenetic, and one of so marvellous a character that we are made to realize, as by a flash of light, how far we still are from a solution of this problem. It may be true, as Schwann himself urged, that the adaptive power of living beings differs in degree only, not in kind, from that of unorganized bodies. It is true that we may trace in organic nature long and finely graduated series leading upward from the lower to the higher forms, and we must believe that the wonderful adaptive manifestations of the more complex forms have been derived from simpler conditions through the progressive operation of natural causes. But when all these admissions are made, and when the conserving action of natural selection is in the fullest degree recognized, we cannot close our eyes to two facts: first, that we are utterly ignorant of the manner in which the idioplasm of the germ-cell can so respond to the play of physical forces upon it as to call forth an adaptive variation; and second, that the study of the cell has on the whole seemed to widen rather than to narrow the enormous gap that separates even the lowest forms of life from the inorganic world.

I am well aware that to many such a conclusion may appear reactionary or even to involve a renunciation of what has been regarded as the ultimate aim of biology. In reply to such a criticism I can only express my conviction that the magnitude of the problem of development, whether ontogenetic or phylogenetic, has been underestimated; and that the progress of science is retarded rather than advanced by a premature attack upon its ultimate problems. Yet the splendid achievements of cell-research in the past twenty years stand as the promise of its possibilities for the future, and we need set no limit to its advance. To Schleiden and Schwann the present standpoint of the cell-theory might well have seemed unattainable. We cannot foretell its future triumphs, nor can we repress the hope that step by step the way may yet be opened to an understanding of inheritance and development.

Reproduced by permission of the publisher

CHAPTER XII

LIFE ON EARTH

FLORA, FAUNA AND ENVIRONMENT

EVERY period and every locality has its particular flora and fauna, and when the environmental conditions change in any way, the forms of life also change. This is a fact which for anyone not ridden by prejudice is practically impossible to deny. Furthermore every living species exhibits those characters which are the most propitious for its continued existence in its particular environment. Aquatic animals assume certain forms and develop certain functions which are quite different from those in the land animals; plants which live in damp surroundings are of a particular pattern more or less, whilst those of the arid deserts are quite different. Temperature, nourishment, warmth from the sun and degree of exposure, the concurrence of other living creatures and many other factors play their part in determining the growth of form and functions in everything that is alive.

It is inevitable that this should be so. If it be impossible to conceive of life without environment, if it be impossible for a creature to live in isolation, independent from all others, if life be an incessant replacement of substance and energy from the environment, the living creature takes from its environment what it needs and gives up thereto what it has a surfeit of; if life is the interrelation between the individual with everything that surrounds it, then it becomes easy to understand such environment as a decisive factor in the development of life. "Organism would be impossible of explanation without an environment" according to Child (1924). Each of the characters of an organism has its relation to external factors. This becomes still more evident when one considers the organism as a whole, which is what it really is. "Unity, order, physiological differences, relationships and the just adaptation of parts among themselves have no significance unless it be in relation to an outside world."

Correlation between form and function in a living creature and the conditions affecting it has always been a cause for wonder in those who study such things. Scientists may differ from one another in opinion in many respects, but the facts, outstandingly clear as they are, admit of no discussion, and though certainly many different explanations may be propounded, reality itself remains indisputable. Even the most radical fixationist theorists have had to admit that each living form is without exception the most suitable for fostering the continued existence of its particular species. According to these latter opinions, this comes about through metaphysical causes implicit in the end of all things.

ADAPTION TO ENVIRONMENT

Jean-Baptiste Pierre Antoine de Monet de Lamarck published his work on Zoological Philosophy in 1809. He attributed the origin of various species in existence to a process of evolution, as had also been done in ancient times by Thales, Heraclitus, Empedocles, Anaximander, Theophrastus, Lucretius, and others. Species continue acquiring their characteristic forms because they become adapted to their surroundings and the modifications which the environment impresses on individuals are then passed on by inheritance. Climate, manner of life, relations with other species are all environmental factors determining the characters of living creatures. Such fresh acquired characters keep on accumulating in influence throughout countless generations under the same environmental conditions, thereby converting the species progressively in accordance with its vital needs. This explains how such characters happen always to be the most expedient to allow a species to live in its corresponding medium. Function, answering to an objective, creates the necessary organ, and disuse on the other hand makes for atrophy. The fresh organ passes along to descendants, in the same way that an organ which degenerates through uselessness will disappear in the course of generations.

According to Lamarckian theory, the individual is active, reacting to its environment, withstanding it or bending before it, and in the last instance becoming adapted thereto by its own labours. Hence it happens, when an individual has eventually acquired the suitable characters needed, that such adaptation becomes naturally a cause for survival. Species, therefore, become adapted, change their form and functions, for the purposes of continuing in existence. Active evolution creates suitable forms and thus preserves life on this planet, despite the changing conditions of our environment in time and in space.

NATURAL SELECTION

Charles Darwin in 1859 explained evolution after a different fashion. Species do not survive because of adaptation, but evolve and change.

From impartial and dispassionate observation it can be shown that they vary with time and according to the place and the influence of various circumstances, or perchance through instability in the species, and such variations become permanent or vanish through natural selection. Fit offspring resist and the unfit become extinct. The same thing occurs in the wild state as may be observed in the breeding of domestic creatures, so carefully studied by Darwin. It amounts to selecting suitable individuals for the purposes of transmission and thus to fix the most suitable characters for getting races which become step by step stronger and more efficient.

Natural selection is merely the possibility, or impossibility, that a species has of living in a particular manner. By evolution and variation of the species, only those will survive which find the conditions necessary for their existence in their more or less immediate world where they are in the course of development. An individual is always affected by surrounding circumstances. It may be that it will disappear in the course of its life through incompatibility of the environment or that it may withstand conditions; obviously if all the individuals disappear then so also does the species, or the line of the species may be so influenced at the end of a few generations that it gradually becomes weaker and weaker to the point of extinction; or, on the other hand, if the environment be favourable, the species will develop and expand in profusion. The environment, the factor of natural selection, and equally the properties of the species will determine the latter's fate. This same natural selection will explain also the origin of those species now extant.

According to the Darwinian theory, environment has no influence on the production of forms. These make their appearance independently from the medium and eventually, according to whether the forms are suitable or not, selection dictates their destiny. It is obvious that the Darwinian idea differs entirely from that of Lamarck. According to Lamarck, the whale takes the form of a fish because it lives in the sea, whilst according to Darwin it lives in the sea because it has the form of a fish.

Darwin does not deny that the surroundings may exert an influence on living creatures, but he does not consider such effect when explaining the production of shapes nor of any other of the biological characteristics of the various species. Genetic factors alone are decisive, and natural selection works on the differences resulting therefrom.

GERM PLASM

After Darwin, opinions of the greatest radicalism came to be expressed, and from this point we get the modern tendency to underestimate the influence of environment on the development of vital phenomena, starting with those of formation. Weismann (1887–92) stated that Nägeli's *idioplasm* or *germ plasm* is found only in germ-cells and that these bear no relation whatever to the somatic cells formed of food-plasm. Through complete isolation of the one type of cell from the others, germ plasm continues from generation to generation, whilst the individual with its body cells plays merely the part of a temporary carrier for the never-dying germ cells. This being so, it becomes plain that even if the individual during its short exist-ence can stand the effect of environment, that is of no account so far as the species is concerned, since the germ cells remain unaffected. This theory is indeed a total negation of the Lamarckian idea of the transmissibility of acquired characters.

Weismann's ideas have gained more and more ground with biologists over the years, including the geneticists themselves, who for the greater part have vigorously championed the theory that the environment is lacking in influence on the individual and consequently upon the species.

Variation, according to Weismann, comes about from *amphimixis*, that is, fertilization or the fusion of two different germ plasms, whereby distinct ids and determinants are blended, and realize between themselves a type of natural selection, say germ selection, which makes the opportunity for certain factors to endure and for others to perish. From these highly complex unions, new forms issue which in their turn are preserved or destroyed by the action of natural selection. Weis-mann is responsible also for the idea that inert determinants, or what we now call genes in the latent state, coexist alongside the active ones. The inerts may come into action through the presence of definite internal or external stimuli. In this way there may come about a struggle between determinants, whereby certain features would prevail and become manifest because their determinants were stronger or more numerous. This would explain how in certain cases there might appear in the progeny, characters which were not evident in previous generations.

SEGREGATION

Some idea of the difficulty attending the separation of the idea of purely genetic action from that of environmental action with its living factors, as if these did not also form part of the environment, can be gained from disquisitions regarding the effects of topographical isolation. Say segregation or isolation fosters the fixation of fresh specific characters, becoming thereby one more means of natural selection, thus preventing crossing with individuals of the original species who do not possess the distinctive characters, then in this way, the newly developed form becomes easily stabilized. This would explain the genesis of odd species through migration or geographical isolation, for instance after some upheaval of nature, or after some remarkable engineering feat which might restrict the availability of large expanses of water or land. Moritz Wagner (1868) maintained that any new and stable form starts with the separation of forms becoming isolated from their originals. He accepts the effects of adaptation and the immediate and primary increment of the modified characters thanks to the crossing of individuals in both sexes which possess them, and the exclusion of those progenitors lacking in them. Romanes in 1886 and Jordan in 1911 also went into this important question, which however contributes but little to the problem of the effect of environment upon the characters of species. We can see, nevertheless, whether any of those propositions deducible from such observations favours more the physiogenic and morphogenic influence of environment through, of course, gene mechanism as suggested, for instance, by le Dantec in 1907.

ADAPTATION AND TRANSMISSIBILITY OF ACQUIRED CHARACTERS

We must now make a distinction between the two related questions: Is there any adaptation of living creatures to their surrounding conditions, or not? and, Are the characters ordained by environment transmitted through heritage?

By reason of Weismann's efforts, many biologists nowadays tend to give a negative answer, though examination of actual cases allows no doubt that life embraces the whole world, since, as Rabaud said in 1922, life does not exist in the abstract, for it consists of living organisms which have to develop within an environment. "Adaptation is no more than the possibility of existence," he goes on to say, "and selection does not actually preserve the best but simply eliminates the most unfit." "The living mechanism," wrote Cuénot in 1925, "has a past which has left its mark, and the aim of every vital activity of a being is to exist and to endure, well or ill, but certainly to survive."

Without arriving at the same conclusion as Jennings, who in 1934 asserted that the same character can be just as much occasioned by genes as by environment, there are some independent naturalists who do not give way before the general current of opinion, and who grant a due morphogenetic and functional value to outside influences. Woodruff in 1941, under the heading "Nature and Nurture" in his *Foundations of Biology*, has this to say, for instance: "Since the life of an organism is one continuous series of reactions with its surroundings, it follows that nurture plays an immensely important part in moulding the individual on the basis of its heritage. Indeed we are apt to overlook the fact, already mentioned, that every character is a product both of factors of the heritage and of the environment and can be reproduced only when both are present. Those characters that appear regularly in successive generations are those whose development depends upon factors always present in normal surroundings. Other characters, potentially present, do not become realized unless the unusual environmental conditions necessary for their development happen to be met. Witness the examples of the supernumerary legs in *Drosophila*, and sun-red in Indian corn. In Drosophila, the abnormal condition of extra legs is inherited in typical Mendelian manner when the flies are reared at low temperatures; whereas supernumerary legs do not appear in flies with the *same* gene heritage when bred at a higher temperature. Or again, the so-called sun-red colour of the kernels of Indian corn is a heritable character that develops only when the kernels themselves are exposed to light by the removal of the husk. In short, the environment, in certain cases at least, may act as a differential intensifying or diminishing gene action, and thus influence the realization of the potentialities of the heritage."

There are strong reasons in favour of this. Surrounding temperatures, barometric pressures, different types of radiation, meteorological and climatic conditions, the metabolic condition, in viviparous creatures, of the mother during the first periods of development of the embryo, all exhibit an influence upon such development, affecting the weight of the offspring, its height or length, its sex and its ability to lead a separate life. This had already been pointed out ages ago by Hippocrates: "Of course the breed may vary even though the seed remain the same, according to whether it be winter or summer, a wet season or a period of drought," and this was again confirmed in 1942 by Petersen and Mayne's lengthy and accurate statistical investigations.

Researches on twins, such as those carried out by Newman, Freeman and Holzinger in 1937, likewise show that environment is the cause of differences even between individuals endowed with the same hereditary potentialities. Not everyone, of course, would be willing to subscribe to the thesis that "what heritage can achieve can also be achieved by environment," though, on the other hand, everybody must admit that the action of environment does get added to the influence of the genes. If it were only the gene factors which were effective in the production of functional qualities or forms, then the only worry of eugenics would be when seeking improvements in human social groupings or in animal and plant stocks. It is easy to show that this in itself is not enough; of equal importance is euthenics, that is instruction in betterment in its most varied aspects.

On the basis of heredity we build up the framework of the *triangle of life*, whose shape and height as well as its area can change considerably, even though its base, heredity, remain constant in all cases. The other two sides of the triangle represent respectively the environment in general and the special case of environmental intervention which is training or education. Though enjoying the same hereditary birthright, the area of the triangle, the capacity of direction in life, may vary considerably.

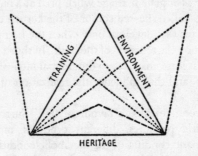

There is no room therefore to forget the circumstances which contribute throughout life, and especially in its earlier periods when the future and the fate of the individual is decided, to the way in which such life shall develop. Conditions in the environment will take part in the production of forms and in the development of functions.

If an individual during its lifetime becomes adapted to environmental circumstances, is it necessary then to discuss the adaptation of species? We find ourselves up against the so well-known question of the transmissibility of acquired characters. We have seen that Lamarck made such transmissibility the keystone of his theory, and that latterly it has been denied practically unanimously.

The material nowadays at our disposal is profuse. We may remember among many others the well-known experiments of Kellog and Bell in 1907 on the silkworm larva *Bombyx mori*. By varying nutrition, functional and morphological modifications were obtained which were transmitted by heredity to a certain number of generations. Tower in 1906 carried out almost simultaneously researches making use of beetles' larva, which he subjected to environments of various degrees of humidity, whilst Picted in 1925 went into the same question, using larva from different sorts of butterflies and moths, which he also caused to vary through feeding and humidity. Variations transmitted from parent to progeny were observed in general even though they always maintained a tendency to revert to their original type.

Cunningham and Cattaneo thought they had found arguments favourable to the thesis of the transmissibility of characters, respectively, in the individual evolution of the flounder, and the humps and horny skin in camels and dromedaries, which characters would presumably be produced by adaptation and be transmitted in series. These arguments are countered by the non-transmission of mutilations, wounds and deformities repeated throughout countless generations, as for instance circumcision amongst the Jews and the bound feet of Chinese women as formerly practised. The docking of the tails of dogs, cats or rats, as effected by Weismann, does not cause the birth of tail-less animals. Against these tests it might be objected that mutilation is not carried out for a sufficient number of generations and that casual excision of an organ of so little biological interest as a tail could not be considered as an acquired character.

Everything rests therefore on determining what exactly is an "acquired character," and when does a particular modification become such an acquired character. Obviously, prior to a character becoming permanent there is no possibility of its being transmitted through inheritance. Normally also such fixation on the genetic potential must be difficult and the influence of environment must be extended through quite a long time for fixation to succeed. Hence it is that investigations in this direction are usually attended by negative results.

It is therefore impossible to state that acquired characters are indisputably transmitted by heredity, even if everything leads to the assumption that such transmissibility may be true. Neither the one side nor the other of this extremely important problem has been definitely proven, and it still remains therefore without a definite solution one way or the other.

MODIFICATIONS, RECOMBINATIONS AND MUTATIONS

We shall say, in short, that individuals may vary through so-called "modifications," through "recombinations" and through "mutations." The first are those changes which supervene by reason of the conditions under which the embryonic or adult life of the individual is developing, these being influences on the body or soma and not on the germ-cells. The environment with all its perquisites, to wit, temperature, humidity, foodstuffs, friends, enemies, etc., as well as education, work, general reaction to the environment, become translated into functional and morphological characteristics such as the blacksmith's brawny arm, the supple fingers of the violinist, the reasoning powers of the mathematician, and so on. These are "acquired characters," and they do not appear to be transmissible.

Recombinations consist in changes in the form of offspring as compared with that of their progenitors, these being exactly proportional to the genetic qualities of such progenitors, and germ characters transmitted definitely in accordance with the laws of heredity.

Mutations give rise to sudden jumps and changes. Normal parents bear children with different forms and functions from their normal selves, occurring by reason of some disturbance of the genetic elements. Mutations constitute an incontestable reason for variation, because the ancestral line will tend to have fresh characters starting with the mutate.

From which we may deduce that an individual can be "modified" merely by the effect of environment, but that fresh characters will not be transmitted by heredity or will be transmitted only with difficulty and perhaps even then only because of some sustained environmental pressure repeatedly exerted upon a large number of generations.

Breeding can be another reason for variation in form or *recombination*. The genetic heritages from male and female parent, which are different when dealing with cross-breeding, will give rise, through Mendelian segregation of characters, to different combinations and the offspring will be perhaps quite different from either or each of its parents. The new characters thus attained will naturally be transmissible. Natural cross-breeds, so effectively studied by Darwin, have had a major part in the generation of fresh specific forms. Crossing, however, which is possible through the cohabitation of two strains then becomes a circumstance arising out of biological environment.

Mutations were formerly considered to be a matter of chance, something mysterious which was produced by unknown means. The transmissibility of mutate characters is now, as we have seen, no longer a matter for discussion. Fresh strains of races and even of species arise from mutation.

We are now beginning to glimpse the mechanism of mutation and it is constantly being shown how influences disturbing the balance of the genes and deficiencies of the most varied nature, physical, various radiations, temperatures, different chemicals, and other actions, are capable of causing mutations by acting on chromosomes and their genes. Mutations are not produced, therefore, spontaneously, but come about from a concatenation of external conditions as part of the environment. Thus we have an acquired character; not necessarily a visible character exhibited by an individual who may have suffered from environmental factors, but it certainly is a fresh character in its germ-cells. Moreover, such character which can cause changes of species is transmitted by inheritance. It is a striking case of hereditary transmission of an acquired character.

Mutation may be of varying degrees of depth and intensity, from the change of an insignificant minor character to a total change of form and function. A slight mutation might resemble an acquired character and this would be transmitted indefinitely. The adaptation might excite the appearance of fresh specific forms through systematic mutation processes consisting of more or less progressive minor mutations. This is the Lamarckian method of variation.

If the abyss between body cells and germ-cells postulated by Weismann does not exist, and if the former are affected by environment and are able to exert their action upon the latter which so plainly do contain genetic powers, and if the germ-cells themselves are also subject to environmental conditions, the problem of transmissibility of acquired characters has lost much of its importance and questions seemingly of the greatest seriousness to us years ago now become mere pastimes. Firm and incontrovertible remains the fact, which nobody can deny, of the fitting of our living world to the qualities of the outer universe in which it exists, both in the spatial as in the temporal sense. Life has therefore to surrender itself without reserve to the circumstances in which it develops.

JEAN BAPTISTE PIERRE ANTOINE DE MONET DE LAMARCK 1744–1829

Philosophie Zoologique (Zoological Philosophy), Paris: Dentu, 1809.

Extensive changes in the conditions surrounding animals will lead to wide modifications in their requirements and perforce to corresponding changes in their activities. Now, if the new requirements become permanent, the creatures then acquire fresh habits which also become fixed like the requirements evoking them. It is obvious, therefore, that a deep-reaching change in surrounding conditions on making itself felt by any particular race of animals will induce in them fresh habits which will, in turn, entail the use of certain parts of the body in preference to those previously in use and in a similar manner will lead to the disuse of any part no longer necessary. All this is known fact and not mere theory.

Now certain well-known facts show that fresh requirements which make any particular part essential will by reason of repeated effort bring about the formation of such part, and consequently it will gradually increase in strength and development and end up by reaching an appreciable size. Moreover, when fresh conditions and requirements render a part entirely useless, the creature will cease to use it and because of such lack of exercise the part in question will gradually cease to share in the development enjoyed by the other portions of the animal, whilst if such conditions of disuse persist over a long period, it will finish up by completely disappearing.

Among plants, where there are no active movements and consequently no habits properly so-called, extraordinary changes in conditions nevertheless also produce wide differences in the development of parts, since in these living organisms everything happens by reason of changes occurring in nutrition, absorption and transpiration, in the quantities of heat and light, air and moisture which they normally obtain, and finally by reason of the ascendance which certain vital motions are able to acquire over others. If such conditions endure, the state of those individuals which are ill-fed, sickly or weak becomes habitual and permanent, whilst their internal organization becomes modified to such a degree that their offspring also preserve the acquired changes and end up by producing a race different from that of such individuals lucky enough always to live in conditions favourable to their development.

What Nature achieves over a long period can be done by man in a day by suddenly altering the conditions in which a plant and all the individuals of a species are situated. Botanists know that plants transplanted from their native heaths to nurseries undergo such profound changes that they may become quite unrecognizable. Cultivated plants become in this way so different from the wild specimens of the same species, that botanists are loth to identify garden plants. Where can we find wild cabbage or lettuce identical with those grown in our market-gardens? How many new strains of domestic pigeon and fowl which it would have been impossible to find in the wild state, have been obtained by breeding them under varying conditions?

However, even discounting changes of site and conditions, we know that living creatures, although they may never stir from where they always have been, are also exposed to outside influences, though the consequent changes may only become apparent very, very gradually. Places change in exposure, climate, nature and local conditions, though with such slowness in comparison with our length of life that we are liable to credit such conditions with being perfectly stable and unchangeable. Now, when conditions change, they exert fresh influences on living creatures, which latter consequently change their habits and become modified, as previously indicated, so as to comply with such alterations of environment. They adapt themselves gradually to the altered circumstances.

HANS DRIESCH 1867–1941

The Science and Philosophy of the Organism, London: Black, 1929.

TRUE FUNCTIONAL ADAPTATION

All other cases of morphological adaptation among animals, and several in the vegetable kingdom as well, belong to our second group of phenomena which, in our analytical discussion, we called adaptations to functional changes that result from the very nature of functioning, and which we shall now call by their ordinary name of "functional adaptations."

It was Roux who first saw the importance of this kind of organic regulation and thought it well to give it a distinctive name. "*By functioning, the organization of organic tissues becomes better adapted for functioning.*" These words describe better than any others what happens. It is well known that the muscles get stronger and stronger the more they are used, and that the same holds for glands, for connective tissues, etc. But in these cases, only quantitative changes come into account. We meet with functional adaptations of a much more complicated and important kind, when for instance, as shown by Babak, the intestine of tadpoles changes enormously in length and thickness according to whether they receive animal or vegetable food, attaining nearly twice the length in the latter case. Besides this, the so-called mechanical adaptations are of the greatest interest.

It has long been known, especially from the discoveries of Schwendener, Julius Wolff, and Roux, that all tissues whose function it is to resist mechanical pressure or mechanical tension possess a minute histological structure specially suitable to their requirements. This is most markedly exhibited in the stems of plants, in the dolphin's tail, and in the lime lamellae in the bones of vertebrates. All these structures, indeed, are such as an engineer might have made who knew the kind of mechanical conditions they would be called upon to meet. Of course all these types of mechanically adapted structure are far from being "mechanically explained," as the verbal expression might perhaps be taken to indicate, and as has indeed sometimes been the opinion of non-discriminating writers. These structures exist *for* mechanics, not *by* it. On the other hand, all these structures which we have called mechanically "adapted" ones, are far from being mechanical "adaptations," in our meaning of the word, simply because they are "adapted." Many of them, indeed, exist prior to any functioning; they are for the most part truly inherited if for once we may be permitted to use that ambiguous word.

However, the merely descriptive facts of mechanical adaptedness having been ascertained, there have now also been discovered real processes of mechanical adaptation. They occur among the static tissues of plants, though not in that extremely high degree which has sometimes been assumed to exist; they occur also in a very high degree of perfection in the connective tissues, in the muscles, and in the bone tissues of vertebrates. Here indeed it has proved possible to change the specific structure of the tissue by changing the mechanical conditions which were to be withstood, and it is in cases of healing broken bones that these phenomena have acquired a great deal of importance both theoretically and practically; new joints also, which may arise through force of circumstances, correspond mechanically to their newly created mechanical functions.

We now have to ask ourselves, of course, if any more intimate analysis of these facts is possible, and indeed we easily find out that here also, as in the first of our groups of morphological adaptations, there are always individual agents of the environment which might be called "causes" or "means" of adaptive effects, the word "environment" being taken to embrace everything that is external to the reacting cells. Of course here also the proof that there are single formative agents does not in the least detract from the adaptive character of the reaction itself. We may perhaps say then that localized pressure is the formative stimulus for the secretion of skeletal material at a particular point of the bone tissue, or of the fibres of the connective tissue, whilst merely quantitative adaptations of the muscles might even allow of a still simpler explanation. Adaptations, nevertheless, remain adaptations in spite of that, even if they deserve only the name of "primary" regulations.

Reproduced by permission of the publisher

LORANDE LOSS WOODRUFF 1879-1947

Foundations of Biology, New York: Macmillan, 1946.

To epitomize—these facts from genetics, taken in connection with the wealth of data from geographical distribution, the succession of types in the geologic past, and so on, give us the modern background for attempting to form an opinion of the method of evolution. The opinion of most biologists is that natural selection in general is a guiding principle underlying the establishment of the adaptive complexes of organisms. Evolution is the result of mutations, germinal variations largely, though not entirely, independent of environing conditions. Many of these variations give rise to characters which neither increase nor decrease the adaptation of the organism, and consequently are neutral from the standpoint of its survival. With regard to such characters natural selection is essentially inoperative. Other germinal changes occur, some of which produce adaptive and others unadaptive characters, and here natural selection is effective. It may eliminate the unadaptive and leave the adaptive variations and so make possible the survival value of the latter in the struggle for existence. The germ plasm never ceases to experiment or natural selection to discover. Variability affording opportunity for adaptability is expressed in "evolvability," a profoundly significant characteristic of life.

So, it will be noted, this is essentially a clarified view of Darwin's idea of natural selection that has been made possible by recent intensive studies of the intrinsic nature and the origin of variations. Natural selection still affords the most satisfactory explanation of that co-ordinated adaptation which pervades every form of life: it shows how nature can be self-regulating in establishing adaptations. But it is probable—indeed, positive—that there are more factors involved than are dreamt of in our biology.

In the words of Thomson: "The process of evolution from invisible animalcules has a magnificence that cannot be exaggerated. It has been a process in which the time required has been, as it were, of no consideration, in which for many millions of years there has been neither rest nor haste, in which broad foundations have been laid so that a splendid superstructure has been secured, in which, in spite of the disappearance of many masterpieces, there has been a conservation of great gains. It has its outcome in personalities who have discerned its magnificent sweep, who are seeking to understand its factors, who are learning some of its lessons, who cannot cease trying to interpret it. It looks as if Nature were Nature for a purpose." Indeed, Aristotle, in effect, emphasized that the essence of a living being is not protoplasm but purpose. However, this thought carries us beyond the accepted sphere of science into the great fields of philosophy and theology.

CHAPTER XIII

GEOGRAPHY AND PALAEONTOLOGY

LIVING CREATURES ON THE EARTH

THE forms assumed by living matter are countless, and vary, as we have seen, in both space and time. They are not identical at different places on the planet nor in different geological periods.

We have seen that each region of the earth, in the seas or on the land, has its own characteristic fauna and flora. The species living in any particular region are determined by a multitude of circumstances, for instance, humidity, temperature, particular foodstuffs available, and other co-existing and competing types of life, resulting in a balance between the various creatures which live contemporaneously or succeed one another. Thus it is that each country is characterized by its flora and fauna.

The areas of diffusion of the various species are also defined according to the circumstances obtaining in them. Plants generally show a more limited distribution, because it is impossible for them to move about. On the other hand, animals, being possessed of the power of movement, are able to wander to greater or less distances and spread. Even so, there are differences among the animals also, particularly observable in fossils, where we find deep-sea sedentary types or *benthos* living on the ocean bed more than a thousand fathoms down and usually showing themselves only in circumscribed areas which sometimes help us to define a particular geological *facies* or period of the earth's history. Wandering creatures, as for instance *nekton* and *plankton*, have a wider habitat.

Fortuitous chances of geography may impede or, on the contrary, foster the spreading of a species. Such particularly is the effect of mountains, streams and oceans. A species at a particular place tends to propagate itself, and the success or failure of its efforts will depend on its own special qualities, which include, for instance, greater or less incentive to live, adaptability, strength, reproductive powers, and so on, whilst on the other hand environmental conditions, be they favourable or adverse to the species, will also play their part. Among environmental conditions we must place foremost the possible existence therein of other living species likely to compete with the particular one in question.

Plants and animals differ according to whether they live in water or on land. Among the former we have marine, lake-dwelling and river forms, whilst on land existing forms vary according to altitude, climate, humidity, and similar factors, and according to whether they live underground or can travel through the air.

The temperature of the sea, its depth, and in connection with this the degree of bathymetric pressure and light, saltiness, osmotic pressure, and chemical composition, nearness to or distance from the coast, ocean currents, presence of certain plants, and particularly of animals, etc., will be conditions which make possible the establishment and survival of any particular species at a particular spot. Among sea and river creatures, certain species remain fixed and others migrate, which latter at certain seasons of the year undertake long journeys in which practically all the individuals of the species move away. Such emigrations are made of course under the influence of those very conditions which, as we have just seen, distinguish the waters where they live, namely temperature, salinity, other plants and animals, foods, currents, and so forth.

Plants and animals in lakes and rivers differ from those to be found in the sea and they vary also according to natural or artificial features of their surroundings. The shutting-off of stretches of water, salt or soft, through natural events or even because of some engineering feat of man, cutting a reservoir of water off from the rest of the region, have as their consequence more or less remarkable variations in the species populating a neighbourhood.

The same occurs on land. Take, for instance, the poor vegetation in deserts or moorlands, the richer pastures and highlands, and the luxuriance of the jungles, all of which lend their stamp to their particular territories. Climate, richness or poverty of the soil, humidity, neighbourhood of particular species, all will cause the flora to exhibit special characters at any particular point, forming spots in which certain species proper to that region will prevail.

There are also areas widely distant from each other where a particular plant lends itself so well to identical climates, that from that fact a common origin is deducible. Such is the case, for instance, of the walnut family *Juglans* which grow in confined districts well separated from one another in both the eastern and the western hemispheres, in Europe, Asia, and America, whilst in past geological ages from the cretacean to the pliocene, plantations of such trees extended over vast zones more than covering their present habitats.

In the distribution of animal life, identical facts are observable, though, as we have said, animals have greater powers of movement. There are, withal, numerous species special to certain limited localities, such as the Arctic and Antarctic fauna, and that to be found in the temperate, warm and tropical regions of the equatorial zones, all of which exhibit their own peculiarities. We have tigers in India and jaguars in America, lions in Africa as well as camels, whilst in South America we find pumas, llamas, and vicugna, kangaroos in Australia, and so on and so forth.

In numerous other cases, similar species populate different continents, where in the distant past they ranged over extensive territories, which is the reason why their present-day descendants form isolated groups widely separated in certain instances.

The opposite situation is however more frequent—segregation. Isolation will give different characters to animals of the same origin. Darwin's observations in this respect are of historic value. During his famous three-year journey round the world in the *Beagle*, he made some findings of the greatest importance in the Galapagos Islands at some six hundred miles distance from the Pacific coast of South America. He describes how the fauna, which as a whole resembles that of the mainland, embraces nevertheless some species not only different from those of the continent but also morphologically distinct from the same species inhabiting other islands in the same archipelago. He mentions the case of certain starlings, which differ according to whether they are found on Charles Island, Albemarle, James or Chatham. At another place, he writes: "It seems remarkable to me to have discovered thirty-eight plants exclusive to these Galapagos Islands, and not to be found anywhere else on earth."

There are over-ground animals which are able to fly and spread abroad, capable of immensely distant migrations in obedience to definite temporal rhythms; and there are other underground species buried deep down in the bowels of the earth, protozoa, worms, arthropods, fish, batrachians, reptiles, mammals, sthenothermic or sluggish in the main, clinging to damp surroundings and many of them blind. Spelaeological exploration constantly reveals the existence of special fauna and flora for each particular system of caverns.

The seas are variously populated according to their situation and depth, marine climate, latitude, and the in-sea distance from a coastline. Everywhere, one can remark a littoral fauna and flora up to a furlong along the coast, whilst deep-sea fauna and abyssal fauna and flora can also be distinguished. The species constituting these groups depend on the depth, the nature of the sea-bottom, the salt content and temperature of the water around. Microscopic creatures from plankton give us animal plankton, *zooplankton*, and plant or vegetable plankton, *phytoplankton*. Then we get phaeoplankton down to 100 ft below the surface, knephoplankton from about 100 ft down to somewhere about a third of a mile deep, and scotoplankton from there downwards to the uttermost depths. And every one of these groups has its particular and characteristic plant and animal species.

The deep sweet waters also differ from the depths of the sea by the nature of their denizens. In some few lakes, such as Baikal, depths down to a mile can be plumbed. In such waters there are no echinoderms, very few coelenterata and seaworms, plenty of articulates, crustaceans, insects and water-mites or their larvae, some fish and reptiles. In lake-waters in general, life is more sparse than in the sea, and still more so in rivers. Every soft-water region, much more than the limitless sea, displays its plant and animal denizens, from the surface and from the depths.

Understandably then, the study of biological geography is fraught with enthralling interest. Apart from the possibility of direct practical applications, its theoretical importance must also be considered. It provides irrefutable proofs of the adaptation of living forms to their environment.

LIVING CREATURES THROUGHOUT THE AGES

In the same way that species vary in space, that is by geographical distribution, according to their circumstances of existence, obviously they will have varied also during the course of time, since in millenium after millenium the conditions in which they have developed have changed time and again.

The knowledge that each geological stratum is identifiable through the fossil remains therein dates from slightly more than a century ago. The majority of substances forming the earth's crust are ancient sediments deposited in ocean depths or in their land-locked connections such as lakes or seas. In this way the various geological strata have been built up, and thus it is possible to determine with greater or less exactitude the period to which they belong. This is done in the first place by the fossils to be found in each stratum, as William Smith first suggested in 1813 and Lyell subsequently confirmed.

The geological age is recognized by the so-called *facies* of the terrain (Gressty, 1838), which is a combination of mineralogical and chemical lithologic features of the ground with the palaeontological characters of the fossils. These give valuable data regarding the biological conditions of the age and environment, height, climate, depth, temperature, saltness, currents, nature of the surroundings, nutritional capacity, and so forth. From all these details we can ascertain the distribution of creatures in existence at a particular time according to altitude and range, which enables us to make up palaeo-geographical charts, as suggested for instance by Lapparent. The facies and formations which are distinct facies though with certain properties in common, since they originated under analogous conditions whether marine, lacustrine, littoral, continental or inland, etc., allow the chronological determination in scale and knowledge of the changes therein and the consequences of adaptation. The same applies to flora as to fauna, so that alongside palaeozoology we must study also the science of palaeobotany.

Together with the influences of physical environment, we repeat, it is important to bear in mind also the biological environment. The composition of a flora and of a fauna in a particular region affects the conditions of existence of each of the beings constituting the one and the other, so that biological reactions between living creatures in competition take first rank in our considerations.

Each geological era is distinguished by the plants and animals which were in existence at that time. There are difficulties in the way of exactly determining the duration of each of such ages, but we do have certain appreciable data which allow at least an adequate perspective of relative durations, even though it may still not be possible to define precisely the length of each period or the time of any particular occurrence. The figures suggested by different writers vary considerably one from the other.

It may be possible perhaps to calculate the total duration of the period included in our geology, that is the formation of the various layers of the earth, as some two billion years. As Woodruff writes: "Ten thousand years are but a yesterday in the history of our earth." From this assumed two billions, more than half would be taken up by the period of no life.

In the long stretch of time reaching from the Precambrian, starting with the Laurentine period, to the Archaeozoic as far as the Algonquin, we can take it that there must have been some bacteria and fungi as well as primitive algae, and most certainly protozoa in existence, all of them single-celled organisms.

Polyplastid organization naturally must have come later. The mosses and giant ferns led up to the conifers and these to the spermatophytes. The plants first were aquatic, from sea and swamp, in huge quantities at certain periods; afterwards came the land plants.

In the lower palaeozoic the smaller crustaceans, worms and sea creatures abounded, after which came the sea scorpions. In the Devonian and Silurian epochs came numerous fish species. Life was still more marine in nature. In the Carboniferous period it extended to dry land; creeping creatures issued from the waters, amphibians descendant from the ancient sea scorpions, giant crabs and huge insects. The Carboniferous is indeed a period of great fertility, and in the strata corresponding to it we find a huge number of fossil remains. Afterwards came a period of drought, when existence became difficult. Temperature underwent considerable changes and long glacial epochs followed. The abounding vegetation of the swamp-lands of the previous age disappeared at certain points, covered by fresh deposits, compressed and mineralized, and carbon, becoming isolated from the organic principles, formed the coal measures embodying hydrocarbons in a gaseous state.

At the start of the mesozoic period, when climatic conditions improved, life on earth renewed its strength and proliferated, amphibians and, even more so, reptiles taking first place. Huge reptiles made their appearance, some of which remained on land whilst others returned to the sea. It is thought that the mesozoic period may have lasted some two hundred million years; the reptilian era may have extended another eight million decades. It was a period of exuberant vitality and many and vigorous species. In the aftermath, however, the generative impulse seems to have become spent, a period of decadence in life started and totally fresh forms of existence made their appearance. A new era of glacial cold kills the mesozoic fauna and flora, the gigantic reptiles and their kin, leaving only those animal and plant species which are able to resist cold. Serious disturbances within the earth mark the end of the mesozoic era and glacial periods alternate with cycles of tremendous heat until we reach the cainozoic and the commencement of the tertiary period, when temperature becomes stabilized once more and life is renewed with fresh vigour.

In the tertiary period birds appeared, descending from the reptiles, whilst mammals also started to develop. The realm of life on earth becomes consolidated and takes upon itself from that time forth the qualities still ruling generally in our present-day existence. In the primitive quaternary epoch the first men come to life.

THE PLASTICITY OF EXISTENCE

Climatic characters and other conditions of existence in each epoch determine the flora and fauna of that period. Environment moulds the species which become adapted thereto or perish; some endure and others become extinct. Fresh forms are constantly being produced in an uninterrupted and incessant burst of generative activity, either by progressive variation, i.e., orthogeny or by mutation. It is likely that in such geological eras the instability of species may have been greater than is shown in the present with all its changes and present frequency of mutations.

Life runs through a course higher than the individual, through groups and collectively; likewise each organism follows its own course through time. This is due to the action of the genetic principle, to which becomes added the unceasing influence of environment. From the union of these two factors we get not only the individual organism, but also, and this is the more important, the species.

The distribution of living organisms in space and in time is thus proved. If life is impossible without environment, if there be no living organism which can subsist independently from the world surrounding it, there is no miracle in the environment giving evidence of its effects on the course of existence, and on the most varied manifestations of life. Genetic agents alone are insufficient to decide the appearance of shapes and forms and other vital peculiarities.

Variations of germ, orthogeny, mutations, cross-breeding, hybridization and segregation bring about fresh characters which are transmitted by heredity and thus become fixed and permanent. Environment also plays its part in the production of distinctive physiological and morphological characters. The coincidence of internal influences with those from the outside ends up in the existential birthright of the world with all its immense wealth. It is that, moreover, which causes individuals and species to be more or less robust and therefore capable of survival or condemned to premature extinction. A species may become extinct through lack of vigour, through lability, or through its inability to reproduce, and similarly, because the

environment in which it occurs is adverse. All these various influences
add together and determine whether a particular species survives or not.
If it is weak, however slight the adversity of the environment, the species
will fail, and conversely if it be strong it will have greater chances of
survival however antagonistic the environment. The destiny of a
species resides in itself, but also in the surrounding world. Fate simultaneously marks both the being and the circumstance.

* * * * *

ALEXANDER VON HUMBOLDT 1769–1859

Voyage aux Régions équinoxiales du Nouveau Continent (Travels in New
World Tropics), Paris: Librairie Grecque-Latine-Allemande, 1816.

I had in mind two objects in making the journey of which I am now
publishing a detailed record. I wished to spread knowledge of the
countries which I have visited and to collect data suitable for throwing
light upon a science as yet but roughly outlined and which has been
vaguely designated by the terms *natural history, geology,* or *physical
geography.* Of these two objects, the latter seemed to me to be the more
important. I was passionately fond of botany and certain branches of
zoology, and I could indulge the hope that our researches might add
fresh species to those already known; still, as I have ever preferred a
knowledge of the logical sequence of long-known facts to that of
isolated ones however fresh, the discovery of some hitherto unknown
genus seemed to me much less interesting than observation of the
geographic relationships of the plants, on the wanderings of plants *in
community,* and on the extremes of altitude to which their various
families could reach up towards the crest of the mountain ranges of
South America.

The physical sciences are interconnected by those same links which
unite all the phenomena of nature. The classification of species, which
must be taken as the basis of botany, and the study whereof has been
made easier and more attractive through the introduction of logical
methods, is to plant geography what descriptive mineralogy is to the
survey of the rocks making up the outer crust of our globe. In order to
grasp the laws governing these rocks in their strata, to determine the
ages of their successive formations and their similarity in the most

diverse regions, the geologist must first of all know the simple fossils making up the mountain masses and of which the study of mineralogy teaches their distinguishing signs and nomenclature.

The same applies to that branch of natural history dealing with connections between plants, whether amongst themselves or with the soil in which they exist, or with the air they breathe and transform. Advances in plant geography depend to a large extent on those made in descriptive botany, and it would be detrimental to progress in the sciences to raise them to the level of general ideas whilst neglecting the value of individual facts.

I have endeavoured to bring together within a single framework the sum total of the physical phenomena offered by that part of the New World lying within the torrid zone, from the Pacific coastline up to the topmost peak of the Andes, that is to say, to combine all the facts about plant life, animals, geological relationships, soil cultivation, air temperatures, the limits of the eternal snows, the chemical composition of the atmosphere, its electric potential, barometric pressure, decline in gravitational pull, depth of blue in the sky, fading of light during its journey through the various layers of the upper atmosphere, the horizontal bending of light, and the heat of boiling water at various altitudes. Fourteen scales arranged alongside a cross-section representing the Andes show the variations sustained by these phenomena through the effect of the height of the ground above sea level. Each group of plants is shown at the height assigned to it by Nature, and the prodigious variety of their forms can be traced from regions of palm and fern up to those of the fever-repellent Joannesia or Chuquiragua, *Juss.*, graminaceae and lichen. These regions form the natural divisions of the vegetable kingdom, and just as the everlasting snows are found at a definite height in any particular climate, so likewise do the febrifuge species such as Cinchona have fixed limits which I have shown on the botanical map attached to this essay on plant geography.

During the course of my tour, I made quite a number of personal notes on the races of man in South America, on the Orinoco settlements, on the hindrances which the climate and vigour of the jungle set in the way of the progress of civilization in the torrid zone, on the character of the country in the High Andes as compared with that of the Swiss Alps, on the analogies observable between the rocks of the two hemispheres, on the physical composition of the air in the Tropics, and so forth.

I had left Europe with the firm resolve not to write what is usually called a traveller's tale, but to publish the fruit of my investigations in a purely descriptive form. I set my facts not in the order of their happening, but according to the relationship they bear to each other. Amid the awe-inspiring majesty of Nature, keenly occupied with the stupendous sights she reveals at every step, the traveller is hardly in the mood to jot down in his diaries that which concerns himself alone and the trivial details of his daily life.

Thus the unity of composition which distinguishes good books from those of faulty design cannot always be faithfully respected, in so far as man cannot help but describe in a more lively manner what he has seen with his own eyes, provided the main part of his attention has been fixed less on scientific observation than on the habits of the peoples he meets and the mighty works of Mother Nature. Now, the most faithful portrayal of customs is that which best reveals the affinities between the members of the human race. The character of nature in the raw or of civilized existence is depicted either by the hindrances which rise up before the traveller or by the feelings evoked in him. What we want to see is he himself ceaselessly in contact with the objects surrounding him, and his story becomes the more interesting the more a touch of local colour is diffused over his description of the countryside and of those who dwell there.

CHARLES LYELL 1797-1875

Principles of Geology, London: Murray, 1830.

SUCCESSIVE DEVELOPMENT OF ORGANIC LIFE

In preceding chapters I have considered many of the most popular grounds of opposition to the doctrine that all former changes of the organic and inorganic creation are referable to one uninterrupted succession of physical events governed by the laws of Nature now in operation. Against this doctrine some popular arguments have been derived from the great vicissitudes of the organic creation in times past; I shall therefore proceed to a discussion of such objections, especially as they have been formally advanced in these words by a late distinguished philosopher, Sir Humphry Davy. "It is impossible," he affirms, "to defend the proposition that the present order of things is the ancient and constant order of nature, only modified by existing laws: in those strata which are deepest, and which must, consequently, be supposed to be the earliest, deposited forms even of vegetable life are rare; shells and vegetable remains are found in the next order; the bones of fishes and oviparous reptiles exist in the following class; the remains of birds, with those of the same genera mentioned before, in the next order; those of quadrupeds of extinct species in a still more recent class; and it is only in the loose and slightly consolidated strata of gravel and sand, and which are usually called diluvial formations, that the remains of animals such as now people the globe are found, with others belonging to extinct species. But, in none of these formations, whether called secondary, tertiary or diluvial have the remains of man, or any of his works, been discovered; and whoever dwells upon this subject must be convinced, that the present order of things, and the comparatively recent existence of man as the master of the globe, is as certain as the destruction of a former and a different order and the extinction of a number of living forms which have no types in being. In the oldest secondary strata there are no remains of such animals as now belong to the surface; and in the rocks, which may be regarded as more recently deposited, these remains occur but rarely, and with abundance of extinct species; there seems, as it were, a gradual approach to the present system of things and a succession of destructions and creations preparatory to the existence of man."

In the above passages, the author deduces two important conclusions from geological data: first, that in successive groups of strata from the oldest to the most recent, there is a progressive development of organic life, from the simplest to the most complicated forms; secondly, that man is of comparatively recent origin, and these conclusions he regards as inconsistent with the doctrine "that the present order of things is the ancient and constant order of nature only modified by existing laws."

It is on other grounds that we are entitled to infer that man is, comparatively speaking, of modern origin; and if this be assumed, we may then ask whether his introduction can be considered as a step in a progressive system by which some suppose the organic world advanced slowly from a more simple to a more perfect state.

The principle of adaptation to which I allude is the same as that in force at the present time and which causes genera and species to be distributed heterogeneously through the various regions of the globe, so that they or the races may exhibit contemporaneously characters differing according to climate, be it arctic, temperate, or tropical, which is not merely a question of temperature. From local conditions

we get the fauna and flora appropriate to each territory. And this is
evident in geography, which is space, as it is in geology, which is time.

CHARLES DARWIN 1809–1882

The Origin of Species, London: Murray, 1859.

I have attempted to show that the geological record is extremely
imperfect; that only a small portion of the globe has been geologically
explored with care; that only certain classes of organic beings have
been largely preserved in a fossil state; that the number both of specimens and of species, preserved in our museums, is absolutely as
nothing compared with the number of generations which must have
passed away even during a single formation; that, owing to subsidence
being almost necessary for the accumulation of deposits rich in fossil
species of many kinds, and thick enough to outlast future degradation,
great intervals of time must have elapsed between most of our successive
formations; that there has probably been more extinction during the
period of subsidence, and more variations during the periods of
elevation, and during the latter the record will have been least perfectly
kept; that each single formation has not been continuously deposited;
that the duration of each formation is, probably, short compared with
the average duration of specific forms; that migration has played an
important part in the first appearance of new forms in any one area and
formation; that widely ranging species are those which have varied
most frequently, and have oftenest given rise to new species; that
varieties have at first been local; and lastly, although each species must
have passed through numerous transitional stages, it is probable that the
periods, during which each underwent modification, though many and
long as measured by years, have been short in comparison with the
periods during which each remained in an unchanged condition. These
causes, taken conjointly, will to a large extent explain why—though we
do find many links—we do not find interminable varieties, connecting
together all extinct and existing forms by the finest graduated steps.
It should also be constantly borne in mind that any linking variety
between two forms, which might be found, would be ranked, unless
the whole chain could be perfectly restored, as a new and distinct
species; for it is not pretended that we have any sure criterion by which
species and varieties can be discriminated . . .

The inhabitants of the world at each successive period in its history have beaten their predecessors in the race for life, and are, therefore, higher in the scale, and their structure has generally become more specialized; and this may account for the common belief held by so many palaeontologists that organization on the whole has progressed. Extinct and ancient animals resemble to a certain extent the embryos of the more recent animals belonging to the same classes, and this wonderful fact receives a simple explanation according to our views. The succession of the same types of structure within the same areas during the later geological periods ceases to be mysterious, and is intelligible on the principle of inheritance.

All the chief laws of palaeontology plainly proclaim, as it seems to me, that species have been produced by ordinary generation; old forms having been supplanted by new and improved forms of life, the products of Variation and the Survival of the Fittest.

As the late Edward Forbes often insisted, there is a striking parallelism in the laws of life throughout time and space; the laws governing the succession of forms in past times being nearly the same with those governing at the present time the differences in different areas. We see this in many facts. The endurance of each species and group of species is continuous in time; for the apparent exceptions to the rule are so few, that they may fairly be attributed to our not having as yet discovered in an intermediate deposit certain forms which are absent in it, but which occur both above and below: so in space, it certainly is the general rule that the area inhabited by a single species, or by a group of species, is continuous, and the exceptions, which are not rare, may, as I have attempted to show, be accounted for by former migrations under different circumstances, or through occasional means of transport, or by the species having become extinct in the intermediate tracts. Both in time and space, species and groups of species have their points of maximum development. Groups of species, living during the same period of time, or living within the same area, are often characterized by trifling features in common, as of sculpture or colour. In looking to the long succession of past ages, as in now looking to distant provinces throughout the world, we find that species in certain classes differ little from each other, whilst those in another class, or only in a different section of the same order, differ greatly from each other. In both time and space, the lowly organized members of each class generally change less than the highly organized; but there are in both cases marked exceptions to the rule. According to our theory, these several relations throughout time and space are intelligible; for whether we look to the allied forms of life, which have changed during successive ages, or to those which have changed after having migrated into distant quarters, in both cases they are connected by the same bond of ordinary genera-tion; in both cases the laws of variation have been the same, and modifications have been accumulated by the same means of natural selection.

ALFRED RUSSEL WALLACE 1823–1913

The Geographical Distribution of Animals, London: Macmillan, 1876.

My object has been to show the important bearing of researches into the natural history of every part of the world, upon the study of its past history. An accurate knowledge of any groups of birds or of insects and of their geographical distribution, may enable us to map out the islands and continents of a former epoch—the amount of difference that exists between the animals of adjacent districts being closely related to preceding geological changes. By the collection of such minute facts, alone, can we hope to fill up a great gap in the past history of the earth as revealed by geology, and obtain some indications of the existence of those ancient lands which now lie buried beneath the ocean, and have left us nothing but these living records of their former existence . . .

The preceding remarks are all I now venture to offer, on the dis-tinguishing features of the various groups of land-animals as regards their distribution and migrations. They are at best but indications of the various lines of research opened up to us by the study of animals from the geographical point of view, and by looking upon their range in space and time as an important portion of the earth's history. Much work has yet to be done before the materials will exist for a complete treatment of the subject in all its branches; and it is the author's hope that his volumes may lead to a more systematic collection and arrange-ment of the necessary facts. At present all public museums and private collections are arranged zoologically. All treatises, monographs and catalogues, also follow, more or less completely, the zoological arrange-ment; and the greatest difficulty the student of geographical distribution has to contend against, is the total absence of geographical collections, and almost total want of complete and comparable local catalogues. Till every well-marked district, every archipelago, and every important island, has all its known species of the more important groups of animals catalogued on a uniform plan, and with a uniform nomenclature, a thoroughly satisfactory account of the Geographical Distribution of Animals will not be possible. But more than this is wanted. Many of the most curious relations between animal forms and their habitats, are entirely unnoticed, owing to the productions of the same locality *never* being associated in our museums and collections. A few such relations have been brought to light by modern scientific travellers, but many more remain to be discovered; and there is probably no fresher and more productive field still unexplored in Natural History. Most of these curious and suggestive relations are to be found in the productions of islands, as compared with each other, or with the continents of which they form appendages; but these can never be properly studied, or even discovered, unless they are visibly grouped together. When the birds, the more conspicuous families of insects, and the land-shells of islands, are kept together so as to be readily compared with similar associations from the adjacent continents or other islands, it is believed that in almost every case there will be found to be peculiarities of form or colour running through widely different groups, and strictly indicative of local or geographical influences. Some of these coincident variations have been alluded to in various parts of this work, but they have never been systematically investigated. They constitute an unworked mine of wealth for the enterprising explorer; and they may not improbably lead to the discovery of some of the hidden laws (supplementary to Natural Selection), which seem to be required, in order to account for many of the external characteristics of animals.

In concluding his task, the author ventures to suggest, that naturalists who are disposed to turn aside from the beaten track of research, may find in the line of study here suggested a new and interesting pursuit, not inferior in attractions to the lofty heights of transcendental anatomy, or the bewildering mazes of modern classification. And it is a study which will surely lead them to an increased appreciation of the beauty and harmony of nature, and to a fuller comprehension of the complex relations and mutual interdependence, which link together every animal and vegetable form, with the ever-changing earth which supports them, into one grand organic whole.

CHAPTER XIV

CAUSATION AND DESIGN

FINALITY AND DETERMINISM

EVEN the most primitive view of the world impresses on man a conviction of intent. Since man acts from his own knowledge of objects within his understanding, phenomena are attributed to the intervention of "entities," "entelechies," as conceived from the standpoints of human consciousness and will. The primitive, in time and space, populates things with mystical spirits through whose agency the activity of the whole is maintained.

Such anthropomorphic myths observe events and foresee the consequences whereby their eventual effect will come to pass, finally, in a manner suitable for achieving a particular object in mind. Happenings develop teleologically "in order" to arrive at a particular result, in physical nature and still more so in the more complex sphere of living things, thus causing Aristotle (to mention but one instance) to say "life is not form but objective."

This proves to be no obstacle against even Aristotle himself, among so many other active minds, to assert that knowledge must be based upon the study of tangible events, on a basis of observation and experience, in order to establish by inductive reasoning the *causes* of such events. The profound ideological revolution which such a proposition signifies should be noted; namely, the endeavour to explain the world by the world itself without the help of myths. Hence we cast aside the symbol of Prometheus who wrested the secret of fire from the gods, attempting to eliminate imaginary deities in favour of an actual explanation of things as they enter our perceptions. If when we achieve some act or do some deed, it comes about through our own will, arbitrarily, and we can likewise not do it, or omit to do it, and if in connection with such act we observe some particular effect which does not occur unless that act be done, we cannot fail to conclude that such deed or act is indeed the cause of the effect, that one occurrence conditions and decides a second, and hence that such *causal relationship* occurs in acts depending upon ourselves just the same as in those which are outside us.

"Show the cause, and you show the effect." "Do away with the cause and the effect is absent." "Vary the cause and the effect is varied." These are Bacon's laws of determinism. An occurrence makes its appearance and develops because it has been preceded, or is accompanied, by other events which are its cause. And Nature thus becomes a continuate and vast concatenation of causes and effects stretching to infinity. Mechanical and physical analysis in all its simplicity demonstrates the relations between causes and effects and hence we arrive at the conception of practical science.

Once indeed the conditions under which events develop are known, the recurrence of such events may be secured by reproducing those same conditions. This conclusion is of great importance for mankind and represents an instrument of incommensurable value in action. The mysterious "beings" are not to be handled by the will of man, though this latter can indeed on the other hand reproduce the causal circumstances and bring about the effects desired. Empirical knowledge, founded on observation or experience, is positive knowledge and there is no need in fact to run to empyrean forces for an explanation of events taking place in this world.

Proceeding on these lines, however, we renounce the idea of a high intellectual hierarchy. Thought "in the mechanical plane" restricts the world to immediate reality, every action proceeds from given antecedents in the same world of such reality. Causes and effects succeed one another or take place simultaneously without end and in this manner events continue to evolve and develop.

Clearly, a determinist framework gives man a knowledge of when and how a particular event will take place and consequently endows him also with the option of repeating it. This is positive science. It is not then a long step to apply experimental standards to the study of the sciences relating to existence and living, which thus naturally become determinist.

This has brought inestimable progress in its train. Positive observation and experimentation are possible in biology and its related and derived sciences and, thanks to the application of such a positive criterion, advances in these branches of knowledge—as in many others—have been tremendously valuable. Observation and experiment—"experience" to put it in a single word—are a sign of science in all its phases, and it would in any event be impractical to deny ourselves the use of experimental standards.

Vital phenomena under their extremely diverse guises are studied through analysis of the conditions determining them. In this way, biological knowledge is built up and thus, likewise, does man continue to acquire means of ever wider scope, and more perfect in application, for taking the measure of vital reality. The standard of causality has enabled its proofs to be definite in the domains of the biological sciences just as in all other branches of positive scientific knowledge.

CONTINGENT REMAINDERS OF PHENOMENA

What we have just said is by no means the whole story. There remains in every science a residue of phenomena, and in a science so advanced as physics we must reckon with Heisenberg's so-called "Uncertainty Principle," the result of which is that some aurhorities argue that the law of cause and effect is at a turning-point in science. From this, it will be understood that there may be in biology numerous facts which seem to elude the concatenation of cause and effect. Throughout the ages vitalistic doctrines of various shades have arisen and endured. We may leave on one side those vitalists who might be said to be so "from impetuosity," because they endeavour to explain *all* vital phenomena by mechanical means and finding this impossible under present-day conditions of the sciences, thunder forth on the inefficiency and breakdown of causal explanations.

There are more serious-minded people who, recognizing that reality is perhaps wider than the frameworks set up by determinism, admit that this does not in any way mean that such frameworks are not efficient and necessary, even though they may be insufficient. Claude Bernard, himself the great protagonist of determinism, wrote: "The great problem dominating everything else is to find out whether logical relationships occurring in events of the mind do indeed tally inevitably and in all cases with external reality, and whether the links we naturally establish between events, for the purpose of satisfying our logical requirements, do indeed really exist. There might be a series of evolutive events following one another in time, which need not necessarily be determinative; a chain in which each link has no cause-and-effect connection either with the one preceding it or with the one coming after it. We shall never reach absolute determinism for everything. There will always be indeterminism in the sciences. But the intellectual conquest of man consists in lessening and rejecting indeterminism as, with the aid of experimental method, determinism gains ground. Only by this can he satisfy his ambition because it is thus that he extends and will extend step and step his power over nature."

TELEOLOGICAL ARGUMENTS

The opinions of the majority of vitalists were always based upon the fact that activities of living organisms respond to possible future conditions. Everything in life seems to respond to some aim of value to that life itself, and this teleology is difficult to square with a purely causal determinism. This is a matter of those "final causes" which so infuriated Francis Bacon. Assumption that an event may answer to a cause arousing it, *pushing* it from behind, as to a motive conditioning it by *attraction*, is what motivates it with a determinate object. This is precisely the working method of consciousness and will to which we have previously referred; human deeds take place in order to attain a foreseen result.

The study of biology is impossible without taking into account the teleological sense of the phenomena observable. This is an argument dear to vitalists of every class. We may quote, among so many of them, von Uexküll's dictum repeated even in 1936 by Noltenius: "Nature, the world of things, has three *dimensions*, namely 'substance,' 'soul' and 'spirit,' so that life may thus be 'material,' 'vital' and 'organismic' or 'systematic.' In the sphere of the material, the dominating motive of events is the 'causal' connection, whilst in the domains of soul and spirit the prevailing action is finality, the determinate end-in-view or *purpose*, a means of connection of a special order characterized by the active influence of the present upon the as yet unfulfilled future."

This extremely simple framework does not seem acceptable just as it stands, but it is as well to point it out as a typical example of the way of thought adopted by numerous biologists, many of them quite eminent, at the present time. Bertalanffy, on the other hand, wrote in 1932: "If the maintenance and renewal of an organism, reproduction, shows that such organism is a functioning whole *with a purpose*, consideration of the purpose will necessarily entail consideration of the complete whole."

FITNESS OF FUNCTIONS

For a long time now the present writer, in *Functional Unity* (1919) and numerous other publications from 1911 onwards, has been pointing out that all functions appear to develop as if they were answering to some particular purpose. Our analysis commenced with the study of the dynamics of the digestive apparatus, after Pavlov had considered working conditions in the glands forming that system, and it then extended to cover all the functions. It is constantly to be observed that if we remain satisfied with purely teleological explanations, physiology of the organs would without exception give us every ground for satisfaction. If we thought that by saying "such function is fulfilled in this manner for such purpose," we had arrived at some fresh knowledge, study would become very easy for us, for it is certainly true in fact that the functions, both individually and in combination as a whole, do seem to obey an intended object.

Of course there is no "archeus" dwelling in the organs to ensure that the working thereof shall be effected in the best possible manner, but on the other hand there is indeed a combination of functional inter-relations from which the appropriate working of the organism is brought about, or, we might perhaps say, an archeus implicit in the event itself.

Completely automatic functions of the viscera develop as if in answer to some definite call, nervous functions still within the realm of physiology, the unconscious workings of the mind—every single one of them acts as if moved by teleological *beings*. This happens not only among a few more highly differentiated animals, man for instance, but is repeated throughout the length and breadth of the animal kingdom. The more primitive forms of life, the more complex, and those higher in the scale, all function as if in their actions they were responding to some final cause. This unquestionable fact lends yeoman support to those who argue that fatalism, the standard of causality, is not always applicable when the phenomena of existence have to be explained.

This biological *adequatio rei*, this making of things fit their purpose, was a discovery of physiologists and biologists. Everything in each and every organism comes about in the best possible way for the fulfilment of its functions, as if a conscient intelligence were presiding over the development of events from which life results. A sense of *use* rules in the world of the conscious as well as in that of physiology. A particular thing, done for a purpose because it is *useful*, is the teleological criterion.

CONSCIOUS AND UNCONSCIOUS INTENTIONALITY

Modern philosophy is generally characterized by its being founded upon the notion of *intentionality*, especially in Scheler and Husserl when reference is made to *vital* intentionality prior to cognition. The primordial emotive impulse—without cognition, feeling or image—is a "leaning towards," or a "deviation from," an adaptation in fact according to need. In animal life initial impulses to action arise from existence itself being directed and governed from that source. "In the higher animals," Scheler writes, "the stimuli exerted on the brain by the endocrine glands represent the more primitive feelings and *sensations* constituting the basis of organic sensations as well as of sensations proceeding from external processes." This is the same idea which was sustained many years before him by the Catalan philosopher Turró, starting from a basis of purely biological speculation. And it is fitting to refer to it here since the general "urge" or "thrust" of growth and reproduction, the "affective impulse," is to be found at the bottom of all vital activities and of all phases of their development, being the direct motive or cause of their *expression*, which meaning was glimpsed at by Darwin as the primordial phenomenon of existence. Hence is derived the instinct and antecedently the successive strata of psychic existence. The conditioned reflex is a teleological act, just as is the whole field of the mental processes.

Scheler opposed "mind" to "existence," making a distinction between process or performance and the "centre of activities" in the person, distinguishing the mind from the functional centres of existence, the mind being, according to him, the focus of the forces combating the pressure of the organic and everything pertaining to life as such. Such life, the psychophysiological condition of an animal, is the starting-point of behaviour in relation to environment. Behaviour, then, is adjusted to vital needs so that life may become possible. The mind directs man's deeds. Behaviour governed by the mind may be opposed to his conduct in life, but even when the results of one act or another happen to be opposed one to the other, in each case they nevertheless obey some standard of utility. Divergence would occur from the different nature of these two types of existence.

An animal cannot carry out this peculiar removal to a distance and substantivation which turns an "environment" into a "universe," any more than it can emotively and impulsively convert the definite centres of resistance into an "objective." It is encrusted and immersed in the vital reality corresponding to its organic conditions, without ever objectively learning anything about the latter. The mind shrinks into itself, becoming conscious of itself, converting resistances into "objects" and environment into a "world." Our impulses, our impulsive life, our original vital impulse precedes all cognition. Awareness of reality is not, therefore, subsequent to but prior to every portrayal of the world. That, however, always implies a movement of desire or of repulsion, a directed activity in all cases with utilitarian

ends and purposes, an "intentionality," which not only must be unconscious but which must also in the majority of cases develop properly
outside or beneath the consciousness.

"The living body is a system of opportunities seeking fulfilment.
The end of every process is its goal. Youthful zest for life and for the
world, to fulfil remote wishes and attain distant aims, that is the vital
pursuance of an objective," as Jung might still well say.

Philosophers, however, generally ascribe intentionality exclusively
to the subjective sphere. Things whose object is known are done on
purpose. "Between the plane of action," says Bergson, for instance,
"and the plane of pure memory, in which the mind preserves in full
detail the picture of past existence, a thousand different planes of
consciousness are to be seen within the aggregate of lived-through
experience. Portrayal of an idea consists in a controlled movement
between the image and the action. The psychic mind will become a
neutralized consciousness and will therefore become latent, ending up
in an individual consciousness. The advance of living matter consists
in a differentiation leading to formation first and to a gradual complication afterwards of a nervous system able to lead excitations through
set channels, and to organize actions. The greater the development of
the nerve centres the more numerous will be the motive routes among
which a single excitation will be able to set in motion various alternative possibilities at choice. Such internal indeterminacy will have,
however, to pass through the mesh of necessity. Freedom thrusts her
roots deep in necessity and is organized in close intimacy with it."

The life of the consciousness is intentionality and nothing else.
Everything done with knowledge and volition is done with a purpose.
Brentano thought that the difference between physical and psychic
phenomena was the intentionality of the latter. "Intentionality shows
itself only in and is characteristic of the acts of consciousness." Again,
according to Husserl, intentionality is the essence of consciousness and
consciousness is the objective of phenomenology. The subject is not
something which may first exist and then subsequently become directed
on to the object. Intentionality is not an *internal* phenomenon. By its
agency, consciousness stands in direct contact with its "world," and
resolves itself into the being existent for the purpose of subjectivity in
the concrete fulness of life.

Heidegger when stating that "heedfulness" is the essence of living,
being the essence of existence, identifying heedfulness with existences
exhibiting intent, did not however imagine that such intentionality

was a mere characteristic of acts of consciousness but that it would explain existence in its entirety, thus agreeing with the conclusions of the biologists previously reviewed.

Over the stream of twenty-three centuries Aristotle still has his say on biological standards of judgment. In our times, Thomson says in 1925: "It looks as if Nature were Nature for a purpose." We may bring Aristotle up to date by stating that the essence of every living being is not protoplasm but purpose.

LIFE AND INTENT

There are nowadays a fair number of biologists who do not admit the validity of the thesis that causality is the giver of shape to vital phenomena, and it is curious to see how physiologists for the most part are mechanists, whilst zoologists, botanists, and naturalists tend more frequently to teleological explanations. The explanation of this is nevertheless quite simple: the physiologists built up their science by means of experiment, biologists by observation. To experiment is to divide reality into small sections of space and time, and in such limited fields to investigate the circumstances and conditions of phenomena and bring about the repetition or reappearance of such occurrences, all of which is impossible without some causal import. Naturalists, on the contrary, freely observe the course of life on the big scale from an entirely different viewpoint and in another perspective presenting a distinctly different picture.

It is a question now of finding out whether these two views of our world are incompatible; whether determinism and teleology are mutually exclusive or not. Thus it was frequently assumed, though doubtless quite fittingly, that there is some synthetical theory embodying both these contrary opinions. Determinism assumes order and mutual dependence of events—nothing takes place in the world unless conditions are right. That is the last word! Intent or purpose is not possible without some order. How could anything happen unless there were an intermediate line of chances along which one must run until the end is attained? This, in the long run, is the same road covered in the same direction, onwards, along certain stages, whose destination in the first case (causality) is unknown to us and of which in the second case (intentionality) the goal is but imperfectly conjectured.

Who can doubt at the present time that in order to know the development of an event, it is necessary to investigate its determinant conditions? Not even the most hidebound teleologist would dare deny this! And conversely, who can dispute that events in life are developed consciously or unconsciously in such a way as if they tended to a particular end? An elementary observation of reality suffices to prove in all cases the exact functional fitness of certain vital manifestations.

Everything appears to pursue one definite object: to exist! Every live creature seems animated by a single purpose: to preserve itself! Life appears effectively to show one purpose in the end: to be perpetuated! The living world runs along certain tracks leading it to a terminus, whilst the engine pulling it on its way is intentionality: a hidden and transcendental purpose whose final designs escape our grasp, and therefore, when the mind of man discovers itself, reflecting upon itself, imbibing knowledge of the universe in its own world of existence, at that magic moment when the mind becomes conscious, universal purpose is reflected in the life of the individual, and like some diaphanous aura from it the voluntary deed is seen as in a glass darkly —"done because it is known to what such act must lead." It is not nature which intentionally works in a similar way to man. It is, on the contrary, man who works with purpose because the events of the Universe are developed on purpose—and independently from man, whose existence is not really essential. Man, a sheer accident, is drawn by an insuperable and transcendental stream towards his fate!

* * * * *

ARISTOTLE 384–322 B.C.

ΠΕΡΙ ΨΥΧΗΣ (Regarding the Mind), said to have been written about 353 B.C.; Greek text—Leipzig: Teubner, 1831.

THE MENTAL FACULTIES

Of the mental faculties previously mentioned, some creatures possess all, some only a few of them, and others one alone. These faculties to which we have referred are those relating to growth and sustenance, perception, movement and thought. Plants possess only the former, growth and sustenance, whilst other living creatures possess perception as well. Any creature endowed with perception must also have needs, for need implies desire, passion and volition. Now all creatures possess at least one sense, that of touch, and anything that can feel may experience pleasure and pain, and where these are to be found there must be desire, that is to say a craving for something pleasurable. Moreover, all creatures experience the need for food, for the sense of touch is a searching for satisfaction. All living beings are satisfied by food either solid or liquid, and warm or cold, which qualities are discovered by the sense of touch. Other qualities are found out only indirectly by contact. Sounds, colours, or smells in no way satisfy hunger; whilst as regards taste that comes within the realm of the tangible. Hunger and thirst are types of need, hunger for what is solid or warm, thirst for what is liquid or cooling, whilst taste is an adjunct to both. These points must be decided later, but for the time being, it may suffice to say that creatures possessing the sense of touch are also subject to needs. As to the imagination, we cannot clearly define it and must discuss it separately. Some types of creatures possess, moreover, the power of movement; others, such as Man or beings like him or perhaps even superior to him, possess also mind and the power of thought.

Obviously, therefore, we may define the idea of mind only in a like manner to shape, since as there is no shape which is not dependent on the triangle and its compounds, neither can there be mind other than in the forms above described. As a matter of fact, with regard to shape we may frame a broad notion applicable to all shapes, though not specifically identical with any particular one. The same applies to the forms of the mind. Thus it is unreasonable to expect a blanket definition either here or elsewhere to cover everything in existence, or on the contrary, in the lack of such overall definition, to try and define each particular species.

Some similarity does in fact subsist between the principles governing shape and mind, since in each case, the outcome implies the origin; thus, for instance, a square implicitly contains triangles, and perception implies needs requiring satisfaction.

The Faculty of Growth

First of all, therefore, we have to consider the questions of sustenance and development, since the growing mind is an element of all other forms of the mind and is indeed the most embracing faculty by virtue whereof any entity may be said to possess life, evidencing itself by the consumption of food and the process of reproduction. The most natural act of any undamaged and fully developed living creature, not spontaneously generated, is the production of other beings like itself;

thus an animal will reproduce itself in an animal, and a plant in a plant, and in this way as far as it is possible they stand their share in the everlasting and supernatural groping towards this end by the very essence of their nature. These words, "this end," are however double in meaning, one interpretation being "the end to be attained," and the other "the end attainable by reason of the act." Now no one living creature can share without interruption in the everlasting and supernatural, for nothing mortal can remain for ever exactly identical with itself throughout the whole course of its existence, whence each plays its part as far as is possible to it with varying proficiency according to its powers. Thus any particular being continues its course not identically in its original form but in some modified form still similar thereto, not in its own original single body, but in other bodies produced from it.

The mind is the origin and mainspring of the living creature, which terms are susceptible to many interpretations. The mind is however the origin for three definite reasons, namely: (a) it is the source of movement, (b) it is "this end," and (c) it is the ruling factor of the living creature. Obviously it is the ruling factor, for in everything the ruling factor is its reason for being, and in living beings that reason is their existence its very self, and the origin and mainspring of such existence is the mind. Moreover, such reality as there is in the latent flows from the conception thereof. Clearly the mind is also "this end," since nature acts in the same way to a purpose in view, and that purpose in view is its final objective or end. In living creatures the mind is of this nature, for all natural bodies are agents of the mind, and whether they be plants or animals it is the mind which governs their existence. It is well to remember here the two senses of "this end," namely (a) "the end to be attained" and (b) "the end attainable by reason of the act." The mind is also the mainspring of movement, though this faculty is not intrinsic in every living thing. Changes in nature and size are also due to the mind, whilst perception, too, may be said to be a type of change which cannot be effected unless a mind is present. This also applies to growth and disintegration, since nothing grows nor decays in the physical sense unless it takes in sustenance, and nothing acquires sustenance unless it be subject to some facet of a mind. Empedocles, however, is not correct is stating in this connection that growth reaches downwards in plants where they take root because the earth naturally has this bent, and upwards because of the similar natural tendency of fire to strive upwards.

RENÉ DESCARTES 1596–1650

Discours de la Méthode (The Right Road to Reason), Leyden: Jean Maire, 1637.

Just as a profusion of laws often furnishes some pretext for wrong-doing, so that a State is the better regulated when having only a few they are the more strictly respected, so also did I think that, instead of that huge number of precepts whereof logic is composed, the four following would be quite enough, provided that I took the firm and steady resolve never once to fail in keeping to them.

The first was never to take anything for granted that I did not know by evidence to be true; that is, scrupulously to avoid headlong thought-lessness and prejudice, and not to embrace in my judgment anything more than what was so clearly and distinctly presented to my mind, that I should have not the slightest reason to be in doubt about it.

The second, to break down every problem that I considered into as many separate simple elements as might be possible and requisite, the better to solve them.

The third, to organize my thoughts, starting with the simplest objects most easy of discernment, in order to rise gradually step by step to an understanding of the most complex, assuming some arrangement even amongst those which did not naturally fall into sequence one after the other.

And finally, in every case, to make such detailed classifications and such general summaries as to assure myself that nothing relevant had been omitted.

Those long series of quite easy and simple axioms customarily employed by mathematicians in order to arrive at their most difficult proofs, had given me some reason to suppose that everything coming within mankind's range of cognition was similarly interwoven, and if only we refrain from accepting as true anything about which there may be the slightest doubt, and keep also to the necessary sequence in deducing each step from its predecessors, there cannot be any conclusion so remote that it will not eventually be reached nor anything so recondite that it cannot be revealed. I did not have much difficulty therefore in deciding how I ought to commence, since I already knew that it was by way of the simplest and most easily discernible facts; and, considering that among all those who have formerly sought for truth in the sciences, only the mathematicians themselves had been the ones to find some proofs, i.e. a few clear and obvious reasons, I had not the slightest doubt that what I ought to do was to start with the same things that they had put under examination, although I expected no other benefit than that this would accustom my mind to feed upon truths alone and never allow itself to be fobbed off with false premises.

CLAUDE BERNARD 1813–78

Introduction à l'Étude de la Médicine expérimentale (Introduction to the Study of Experimental Medicine), Paris: Baillière, 1865.

A fact means nothing by itself alone. Its value lies solely in the idea intrinsically borne within it or in the evidence it may serve to form. We have said elsewhere, that when a fresh fact is labelled as a "discovery," the discovery is not the fact itself but the new idea which that fact leads to. Similarly, when a fact demonstrates something, the demonstration is not the fact itself, but the logical relation which it helps to establish between the effect and its cause.

The truths of mathematics are absolute and belong to the mind, because the *ideal* or non-substantial conditions of their existence are also mental and perceived by us in an absolute sense. Experimental truths, on the other hand, are relative and not mental, because the *real* substantial conditions of their existence are also non-mental and cannot be known to us except in their relation to other things in the state that science stands at present.

But though the experimental truths which serve as the basis for our reasoning be lavishly enwrapped in the complex reality of natural phenomena, from which it follows that they cannot appear to be anything more than isolated fragments, that does not mean that such truths of experiment are not founded upon principles which are as absolute as those of mathematics, and that like these they are accessible to the mind and to logic.

The absolute principle of experimental science is a *rigorous determinism* of the conditions of the subject, so that, given a natural occurrence, no variation whatsoever shall occur in it without changes having been simultaneously produced in the conditions in which such occurrence comes to pass. It shall not be valid to admit any variation observed not answering to a particular cause. Having now the antecedent *a priori* assurance that each variation is determined by one particular cause, it follows that rigorous and logical relationships between cause and effect may exist in the evolution of the occurrence.

Experience teaches us the form of the phenomena, but the relation of an occurrence to a determinate cause is necessary and independent of experience—it is mathematical and absolute. Thus we reach the conclusion that the *criterion* of experimental science is at rock bottom essentially identical to that of the mathematical sciences, since in each case the principle becomes evident through a similar necessary and absolute relation. What apparently distinguishes the one case from the other is the complexity of the phenomena in the natural sciences as against the inexorability of mathematical reasoning. Natural phenomena are to be found involved in and clouded by other numerous, complex and varied factors which accompany them, and render difficult an inquiry into the causal conditions of the phenomenon under consideration. By trial and error we analyse and disassociate the various phenomena, for the purpose of bringing them down to ever more simple terms and relations, and we hope thus to arrive at the real form of scientific truth, or, in other words, discover the law which allows us to interpret changes in the phenomena.

Such analysis by trial and error, or what we prefer to call experimental analysis, is the only means of finding out truth in the natural sciences, and the *absolute determinism* of its occurrences of which we have the *a priori* knowledge has to be the criterion directing and sustaining us in our investigations.

We find ourselves still very far from an absolute truth, however much the effort we may have made, and it is probable, particularly in the biological sciences, that such truth may never be reached. This must not discourage us by any manner of means, since, despite everything, each step forward in science represents an approximation and because; through experimental research, fresh and closer relationships are becoming established one by one among the natural phenomena and facts, which relationships, partial and relative though they be, allow us to expand our mastery over nature on an ever increasing scale.

From the foregoing, it could be deduced that if an experiment gives us an event apparently at contradictory variance at any point with what is already known, and which does not tally logically with the determined conditions of existence, reason would have to "repudiate the fact" as being non-scientific. It would, however, be necessary first to seek by experiment to find out if some cause of error might perhaps have disturbed or upset our factual observations. Most assuredly, there is bound to be some mistake or insufficiency in the observation. To admit the appearance of an occurrence without valid reason, and without determinative conditions, is the same as denying science itself.

Confronted with a case of this nature, the scientist cannot hesitate for an instant. He must give science his full confidence and throw doubt on the other hand upon his qualifications as an investigator. He will have to perfect his methods, and use every effort to find the reason for any apparent contradiction, but it must never enter his head to deny the *determinism* of the phenomena, since it is precisely the feeling and conviction of such determinism which characterizes the true scientist.

Facts are frequently adduced in the practice of medicine, which have been faultily observed and are therefore apparently indeterminate, constituting by the same token veritable stumbling blocks in the way of science. It is said: "That is a fact—it must be admitted." Rational science based on a stringent determinism can under no circumstances evade a properly observed and exact fact. However, for that very reason, it must never be cluttered up with facts gathered inaccurately without discernment and lacking therefore any real meaning. Science rejects the *indeterminate*, and when, for instance, in medicine, somebody attempts to found opinions upon the "magic touch," the "sixth sense," inspiration, or a more or less vague intuition, he will fall unpardonably into a fanciful, arbitrary, anti-scientific and ever dangerous type of practice, since he thus delivers the patient to the whims and imaginings of a fanatical ignoramus. True science knows how to doubt and to refrain from ignorance.

If the physicist and the physiologist are distinguished by the fact that one studies things which occur in inanimate matter whilst the other is concerned with those which happen in living matter, there is no difference between them, on the other hand, as to the object each has in view. Both indeed have *the common objective of investigating the proximate cause of events.* Now what we call the proximate cause of an occurrence is the material or physical condition of the thing in being or the manifestation thereof. The object of the experimental method or the end of any scientific investigation is exactly the same when studying events in inert objects as when dealing with live creatures, and consists in recognizing the relations uniting an event with its proximate cause, in determining the conditions necessary for the event to be produced.

When the experimenter has succeeded in learning the conditions surrounding an event, he becomes, so to speak, master of such event. He can cause it, prevent it, help it, hinder it, predict its course and its forms of appearance. The experimenter will then have gained his object and in this manner, by means of science, have extended his sway over a natural occurrence.

We shall therefore define physiology as "that science the object whereof is the study of events in living creatures and the 'determination' of material conditions for the appearance of such events." Only by the analytical or experimental method shall we be able to arrive at the determination of such conditions, in either lifeless or in living things, for in all sciences logical reasoning remains the same and consequently the idea of experiment is basically and in all fields essentially identical.

PLATO 428–348 B.C.

ΣΥΜΠΟΣΙΟΝ (Symposium: The Dinner Party), said to have been written about 400 B.C.; Greek text—Leipzig: Teubner, 1898.

At this point Pausanias paused—note the alliteration I've caught from our rhymesters—and according to Aristodemus, Aristophanes should have been the next to speak, but either through over-eating or from some other cause, he was suddenly taken with a bout of hiccups which prevented his taking the floor. Turning to Eryximachus, the doctor who was next to him down the table, he managed to murmur: "Look here, Eryx, it's up to you either to stop my hiccups or else take my turn at talking until this fit is over." To which Eryximachus replied: "All right! I'll do both, for not only will I speak in your place, so that you can take my turn when your attack passes off, which will be all the sooner if, while I'm speaking, you hold your breath for a stretch; if that doesn't stop it, you'd better try a bit of gargling with water, and if it still keeps on, tickle your nose with something handy and sneeze once or twice, which will chase the hiccups off however persistent they may be." "Well and good," said Aristophanes, "now get on with your talking, while I start following your prescription." Upon which, Eryximachus commenced, as follows:

"Since Pausanias did not, to my mind, sufficiently develop the theme he introduced with such verve, I think I can finish it off for him. I quite agree with his dividing Love into two kinds, though the Art of Medicine has, I think, taught me that Love does not reside solely in the human mind, but is present in all creatures and in a great variety of other things, working in the bodies of animals and in all the fruits of the Earth, in short in everything that exists, so that its power and marvels hold universal sway over all matters both human and divine.

"In honour of my profession, I shall take Medicine as my first illustration. Now, practically everything has its converse—cold has hot; bitter, sweet; dry, wet; and so on. And it was by knowing how to promote attraction and harmony between opposites that Aesculapius, as our poets say and as I myself verily believe, became the Father of our Craft, and I dare say Love governs Medicine just as it rules Athletics and Agriculture. Music also, not to make a lot of bones about it, is in a like case, and maybe that was what Heraclitus meant by that puzzling phrase which he never bothered to explain more fully, namely: 'Unity acting upon itself becomes automatically regulated and produces Harmony as, for instance, in the twanging of the Bow or the Lyre' . . .

"Well then, if you believe that the natural object of Love is what we have so many times agreed, the question need not disturb you, since now, as in times past, it is mortal nature which ever seeks to save itself as far as possible from extinction and to become eternal, and the only way it can do that is by birth ever producing a new individual in place of the old. Though indeed we may pretend that a person lives and is the same from his birth until his death, in actual fact he never is exactly in the same condition with the same properties, since everything about him is in a constant state of decay and renewal—his hair, his flesh, his bones, his blood, and in short, his whole body; and not only his body, but his soul, his habits, manners and customs, his opinions, wants and desires, his pleasures, pains and fears; none of his feelings remains unaltered, but all are in a constant state of flux, waxing and waning all the time. The most surprising thing is, however, that not only do our faculties and knowledge blossom and wither within us in the same way, since in this connection also we are ceaselessly changing, but that every single one of us passes through the same type of change. Indeed what we call 'meditation' relates to some experience which is fading since forgetfulness is the extinction of something known, so that reflection or pondering is the formation of a fresh recollection to take the place of the one which is vanishing, so as to preserve within ourselves that knowledge which we think is the same as before.

"This is the way in which all mortal beings are preserved; they do not remain absolutely and ever the same, as is the case with divine things, but what moves on and grows old leaves in its place a fresh individual similar to what it was itself. This, my masters, is how everything mortal partakes of immortality, in both its body and in all other respects. It is not to be wondered at, therefore, that every animate being treasures its offspring so much, since the anxiety and love which imbue it has no other ground than this thirst for immortality which is its final aim."

AUGUSTINE OF HIPPO (S. AURELI AUGUSTINI)
A.D. 354–430

Confessionium (Confessions), said to have been written about A.D. 399

In what way, however, can the as yet non-existent future shrink or be consumed, and how can the now no longer existent past grow, unless three processes take place in the mind regarding them? For, the functions of the mind are to consider, to experience, and to remember, and what it considers and experiences passes into what becomes memory. Can it therefore be that what is to happen in the future is not yet in existence, even though there may already be in the mind an anticipation of things to come? Can one deny also that the past is now no longer existent, even though there remain in the mind a memory of things past? Furthermore, can it be denied that the present lacks duration, since it passes instantaneously into the past? Moreover, experience lingers, so that what formerly was to come to pass, becomes something which has happened. It is not, therefore, the as yet non-existent future itself but the anticipation of a distant future which is a "long time," any more than it is the now no longer existing past which is long, but the back-reaching memory thereof.

Say I have a mind to recite a piece I have learnt. Before starting, my consideration extends over the whole of it, but when I have begun, every part of it as I bring it out drops away successively into the past and enters the domain of my memory, and the existence of this action of mine becomes divided into a memory of what I have said and a reflection of what I still have to repeat. My experience remains, however, in the present, through which what is future travels into the past. The further I go with my repetition, the shorter becomes my foreknowledge and the longer becomes my recollection, until the moment when the whole of my reflection is consumed and the completed action passed into memory. What is true of the piece as a whole is so for every single part of it, and for every syllable, and similarly for any longer action of which the piece in question may be but a small portion.

This applies also to the whole course of man's life, which is made up of all his separate actions, and to the whole history of mankind through the ages, the parts whereof are the complete lives of all men put together.

MAX SCHELER 1874–1928

Die Stellung des Menschen im Kosmos (Man's Place in the Universe), Darmstadt: Reichl, 1928.

We have climbed a bit into the sublime. Let us return to the problem of human nature lying nearer to our experience. For the present age, the classical theory of mankind found its most effective form in Descartes' doctrine, which doctrine we have in fact only quite recently begun to shake off absolutely and entirely. By his classification of all matter into either "conceptual" or "protensive," Descartes introduced into Occidental thought a whole host of most weighty errors regarding human nature. By such division of the entire universe, even he himself had to adopt the absurdity of denying mental qualities to all plants and animals, and interpret the "semblance" òf mind in animals and plants—which in all ages prior to him had been taken for granted —as an anthropopathic "groping" of our vital feelings into the external images of organic nature, as well as to declare purely "mechanical" everything which is not human consciousness and thought. The result of all this was not only to overestimate in a most absurd manner the particular status of man, emancipating him from the very arms of Mother Nature, but also by a stroke of the pen to do away with the basic classification of life and its primary manifestations as far as this world was concerned. The universe, according to Descartes, is composed consequently of nothing more than points of contemplation and a powerful mechanism which must be explored purely from the mathematical standpoint. One thing only in this doctrine is of any real value, and that is the novel independence and sovereignty of the mind, and recognition of its superiority over and above everything organic and merely animate. All the rest is utter nonsense.

Nowadays we are entitled to say that the problem of body and soul that has occupied mankind for so many centuries has, so far as we are concerned, lost it metaphysical significance. The philosophers, doctors and scientists concerned with this question, concur more and more harmoniously on a basic first principle. That there is no such thing as a locally determined incorporeal matter—as postulated by Descartes—is therefore quite self-evident, since neither in the brain-box nor anywhere else in the human body is there a central point where all the sensitive nerve fibres can meet together and all nervous processes combine.

Another radical fault in the Cartesian theory is his premise that the mind consists only in "consciousness" and resides exclusively in the cerebral cortex. Exhaustive research on the part of psychiatrists has proved to us that the decisive mental functions constituting the foundation of the human "character," particularly everything belonging to instinctive activity and the passions, which we have indeed acknowledged as the basic and primal form of the mind, does not in general have its parallel physiological processes in the upper cavity of the skull but in the neighbourhood of the brain-stem, partly in the grey matter of the central cavity of the third ventricle, partly in the thalamus, which acts as a central clearing-house between sensations and impulses. Moreover the ductless gland system (thyroid, genital, pituitary, adrenal glands, and so forth) whose operation determines human passions and emotive life, as well as abnormalities of height and thicksetness, giantism and dwarfism, and probably also racial characteristics, has been shown to be the exclusive seat of regulation between the whole organism including its morphological development and that small appendage to mental activity which we call the waking consciousness. It is nowadays by no means the brain by itself but the whole body which once more becomes the field of physiological parallelism to the mental processes. There can no longer be any serious support for such an external linkage between the mental substance and the corporeal, as assumed by Descartes. It is one and the same life that possesses mental form in its "subjective being" and bodily shape in its objective relationships. We may not adduce against this, that the "ego" is therefore simple and unique, whilst the body on the other hand is a complex "community of cells." Present-day physiology has completely destroyed the idea of a cellular community, just as it has also broken with the basic idea that the workings of the nervous system combine merely by summation of parts, that is to say not in a ready-made integrated whole, and at times are rigorously determined in location and morphologically in their points of departure. If, indeed, like Descartes we deem the physical organism to be a type of machine, and this in the strict sense of the old mechanist philosophy of the age of Galileo and Newton but now quite ousted by physical and chemical theory and cast on the scrap heap; if in addition, like Descartes and his followers, we disregard the independence and the certainly established predominance of the combined emotive instinctive life over all conscious imagery; if we also restrict all mental activity to the waking consciousness, discounting the enormous cleavages between entire co-ordinated groups of functions of mental occurrence and the conscious ego; if, furthermore, we deny any repression of emotions and disregard any conceivable recalling to mind of whole phases of life, such as the well-known disassociation manifestations of the conscious ego itself in schizophrenia—then, to be sure, we arrive at the following false antithesis: whilst on the one hand there is unity and simplicity of a primitive type, on the other we have a multiplicity of bodily parts combined in a secondary manner as well as a plurality of processes primarily founded thereon. Such a picture of the mind is quite as erroneous as the notion of the spontaneous generation of life presented by a more ancient philosophy.

In resolute opposition to all such theories, we may therefore say that the physiological and mental processes of life are rigorously identical from the ontological point of view, as indeed Kant had already surmised. They differ only in appearance, but are apparently rigorously identical in their laws of structure and in the rhythm of their progress through time: both processes are non-mechanical, the physiological just as much as the mental; both are directed to the final aim, teleoclinic and based upon totality. The physiological processes are all the more of this nature the lower (not necessarily the higher) the position of the parts of the nervous system in which they run their course; the mental processes likewise are more complete and directed to their target, the more primitive they happen to be. These two processes are merely two aspects of the single process of existence, which is higher than purely mechanist because of its own configuration and the interplay of its functions. What we thus call "physiological" and "mental" are only two facets from which to view one and the same vital process. There is a "biology from within" and a "biology from without." The latter starts with a perception from the structural form of the organism and progresses to the vital processes properly so-called, but it must never be forgotten that every living form from the ultimate differentiable cell elements through the gamut of cells, tissues, and organs up to the organism as a whole, is at every turn and instant dynamically sustained and renewed by such vital process, and that in development it is the "formative functions," sharply distinct from the operational activities of the organs, which produce the static forms of organic matter thanks only to the assistance of the chemico-physical "state of affairs." The late Heidelberg anatomical specialist Braus and E. Tschermack on physiological grounds, justly took these ideas as the focus for their joint research. It may be said that this same concept nowadays permeates all sciences concerned with this famous problem. The old "psychomechanical parallelism" now verily belongs as much to the scrapheap as the "theory of reciprocal action" as resuscitated by Lotze, or the mediaeval scholastic hypothesis of the soul as a corporeal shape.

The physiological "function" is in its basic conception an independent regulated succession of mobile form, a figure moving through time and from its very beginning in no way rigidly tied down to a particular location but able with much greater freedom to choose and indeed build its field of operation in the available cellular substrates. A cumulative organic reaction of a determinate and rigorous nature does not exist either among those physiological functions which possess no sort of correlation in the conscious; in fact they do not exist, as has recently been proved, even for so simple a reflex as the knee jerk. Apparently, moreover, the physiological behaviour of the organism is quite as "clever" as the conscious processes, and by the same token these can often be just as "silly "as the organic processes.

In my opinion, research should straightway be set the methodical aim of testing in the widest measure how far the same behaviour methods of the organism can be induced and varied in the first place by physico-chemical stimuli from without, and again through mental stimulation—suggestion, hypnosis, all kinds of psychotherapy, changes in social environment—on which illnesses are dependent much more than is suspected. Let us, therefore, definitely guard ourselves against setting too high a false value on exclusively "psychological" interpretations. As we know from experience, a gastric ulcer may be conditioned and governed just as much through the mind as through any particular physico-chemical process; and not only nervous illnesses but also organic indispositions each have entirely definite mental correlatives. Also we can quantitatively evaluate both types of our influence on the intrinsic unitary vital process, to wit through the channel of the consciousness and through that of external bodily stimulation, in such a manner as to husband the one stimulus to the same extent as we squander the others. All these are merely different methods of approach, which we retain in our experience and conduct for one and the same ontologically centralized vital process. Even the highest mental functions, such as so-called "connected thought," do not repudiate a strong physiological parallelism. And finally, in accordance with our theory, the intellectual acts must also possess a physiological and parallel linkage, since they derive their entire active energy from the living sphere of impulse, and cannot without some sort of "energy" manifest themselves for the purposes of our experiments even to the individual experimenter himself.

Reproduced by permission of Prof. Maria Scheler

JOAQUÍM PALAU XIRAU 1895-1946

La Filosofía de Husserl: Una Introducción a la Fenomenología (Husserl's Philosophy: An Introduction to the Study of Phenomenology), Buenos Aires: Losada, 1941.

INTENTIONALITY: VOLITION OR PURPOSE

Since the world of reality depends on cognition, and is established on that as a foundation, then in order to discover the ultimate basis to which we strive, it becomes essential to determine the intrinsic structure of consciousness in all its aspects and limits. For such purpose, it is vital to commence with the concept of *intentionality* or design, which, in line with Brentano's terminology, Husserl ascribes to *the consciousness as its ruling characteristic.* As we saw in the brief survey previously outlined, this fact of design is not intention in the sense of finite purpose or practical directivity of existence. It means simply, that every perception is perception of an object perceived, every thought an idea regarding an object thought about, every wish a want regarding an object desired, every affection a feeling for an object held in esteem, . . . every notion and idea of an object is a notion and idea of an object conceived and meditated over.

From all the foregoing, we gather that Husserl's concept is something decidedly distinct from Brentano's, which is its immediate antecedent, and from the Aristotelean and Scholastic, which for the first time defines *consciousness as a mental intention*.

For Scholastic philosophy and even for Brentano, with certain profound differences which it is needless to enter into at this point, the relation of intent comes to pass between two previously formed and antecedent points of reference. A connection becomes established between subject and object, by means of which reality is presented to the subject against which it sets itself as something externally opposed or *objective* in the etymological sense of the word, and such objective amounts to an understanding and possession of the mind. *Intention is a relation between a psychological act and its content, between a real subject and the content of a real objectivity*. It is thus a property of the consciousness characterizing it and even defining it, but only on the assumption of its existence and the existence of *the object to which it refers*. Among all the substances, in the physiological meaning of the term, one only—consciousness or cognition—possesses the faculty of placing itself in a position of direct contiguity with the remainder of such substances which latter are, by definition, transcendental.

Husserl's intentionality has a much deeper function. A substance, anything whatever from the world of things or what we call *reality*, has its own structure and consistency. The whole existence of things is based upon their mutual and reciprocal exclusion and they enter into relation only when certain private limits are overstepped. Relationships between them are the consequence and are to a certain degree external to them. Cognition has no reality outside the relationship it sets up; its whole existence is exhausted in the mere act of setting itself up in relation to something other than itself. Intentionality is not a relation added to the previous existence of subject and object; it is the primary and governing fact of cognition, wherein and whereby subjectivity and objectivity are produced and brought face to face. The existence of consciousness subsists in intent. Thus intentionality is not a property of cognition, a characteristic following on its existence.

It is something integral in its very existence itself. Furthermore, it is not directed towards something "outside" cognition—as in the realist theory—nor towards some concrete "content" of the conscious—as in the idealist theory. The object on which it is trained is not a "real" element of the conscious, nor a "reality" beyond it, but an *ideal* objectivity. By the act of intent animating it, existence passes beyond itself, becomes transcendental, and acquires a specific form of objectivity.

Intentionality does not signify the road whereby a subject previously existent alongside the object confronts the latter and makes contact with it. It is the very essence and definition of subjectivity. The subject is not something which first exists and consequently thereafter becomes directed on the object. Intentionality is not an "internal" phenomenon. By its means, cognition makes direct contact with the "world," and the world becomes reduced to an entity existing for subjectivity, in the concrete fullness of life. The life of the consciousness is life in the presence of "being."

CHAPTER XV

REFLEXES, CONSCIOUSNESS AND WILL

TROPISM AND TACTISM

EXTERNAL agents of various types, in addition to their own internal impulses, bring about in living organisms a variety of acts which, in general, promote the survival of the individual or species. Such reactions are in their most diverse applications favourable to the development of life.

This is the case with tropisms in particular among very many possible examples. Plants shoot forth their foliage into the air—negative geotropism—and thrust their roots into the ground—positive geotropism—under the influence of gravity. The climbing plants, through the action of thigmotaxis, wind around tree-trunks or other solid objects which serve them as support as they stretch forwards. Another frequent occurrence is phototropism or bending towards the light by plants, and we could mention countless other instances.

Such automatic positioning comes about through the agency of certain factors particularly sensitive to definite impulses, located in the cytoplasmic extensions of the epidermal cells of climbing species, or in the hairs or tactile protuberances of *Drosera* (Sundew), *Dionaea* (Venus's Fly-trap), *Mimosa* (Sensitive Plant), and others among the coleoptiles in general. Material factors of metabolic or hormonic nature may also come into play.

In even the most primitive independent forms in the animal world, and even among plants as well, motor reactions to various stimuli are observable, acts of *taxis* aroused by light, heat, osmotic pressure, electric energy of various kinds, or chemical substances. Movements are steered in one direction or another, approaching or fleeing from the stimulus, according to the nature of the latter and the condition of the organism itself. There is a huge variety of *tactism*, and study thereof has permitted the discovery that movements, even in the rudimentary forms of life, evolve in accordance with a standard of usefulness, as if answering a particular purpose, namely sheer survival.

Phenomena develop in the best possible manner, which caused Binet (1904) and Jennings (1906) to say that, in the protozoa and infusoria, movements come about as if governed by intelligence and will. The most effective stimulant is made up of nutritional matter. Jennings (1897) and Wallengreen (1902) showed that in paramecia, slipper-animalculae, activity increases and sensitivity to excitants, particularly chemical excitants, becomes livelier when the creatures are in a state of hunger, lacking food. Nutritional chemotaxis in amoebae is common knowledge.

The behaviour of protozoa answers to the conditions in which the life of the creature develops. Mast (1910) investigated, for instance, the influence of light on *Euglena*. Jennings (1906) published his famous work on the reactions of the lower animals, studying especially the case of controlled movements in slipper-animalculae. These ciliates evince movements in defence, of trying to get away from an unfavourable stimulus, flight, avoiding reaction, or *taxis*, or, on the contrary, approaching the point of origin of a favourable stimulus, i.e. positive taction. Motions may be organized in such a manner that the creature weaves tentatively around the object which is the cause of the excitation process, in accordance with a system of trial and error, terminating in the adjustment of the living organism to the circumstances of the environment. In this way, by suitable action on the part of the creature, environment contributes to its survival.

The vital, motor, responses of such rudimentary, unicellular organisms, before the incidence of external conditions, are fulfilled as if there were some governing purpose, some intelligence or some will set in action by such intelligence. Facts of this nature have been repeatedly observed in various species of protozoa and protophytes.

When the structure of the living organism becomes complex, which entails intrinsically a parallel increase of complexity in reactions, a like congruence obtains. Behaviour becomes more marked in creatures endowed with movement.

In the spongiaria obvious reactions to definite stimuli are seen, consisting in occlusion of the pores and osculum and contractions of the cells which might obliterate the small cavities of the canalicular system. These reactions may be local or may even extend throughout the sponge, and are followed by a slow return to the original state of rest. The movements are due to specialized contractile cells, without intervention of nerve or sensory cells, which such creatures do not possess. These contractile cells are "independent effectors" responding to exterior stimuli through the irritability of their protoplasm and answering similarly to the chemical conditions of the cells which in combination make up the individual. This transmission of impulses from cell to cell, which is still not yet a nerve transmission, is called "neuroid transmission," promoted by excitations proceeding from the environment and through changes in state of the actual cell plasma of the creature. The movements, in this case also, respond mainly to metabolic conditions.

THE NERVOUS SYSTEM

In hydrozoa, among the coelenterata, nerve elements are to be found, as well as differentiated effectors. Setting aside the cnidoblasts, extremely simple in function, we must count as instruments of motion in the hydra, the muscular arrangements, both round and long, in the body, hypostome, base and tentacles. The contraction of these muscular fibres comes under the control of nerve elements. The ectoderm contains a nerve-net in which at least four classes of sensitive cells are to be found. There are further nerve cells associating among themselves and sending out extensions to the muscular fibres. In the endoderm there reside only a small number of isolated nerve cells and a few others which appear to be sensitive. The ectodermal nerve elements are abundantly in evidence in the hypostome around the stoma, in the base of the tentacles and in the foot. They form a continuous syncytium, the extensions continuing from cell to cell, without break, in contrast to what happens in the case of the nerve centres of more highly differentiated creatures, in which each neurone—as revealed by Cajal—maintains its cytological unity.

By means of the nervous system, not yet centralized, the effects of environmental stimuli of distinct type are developed and, furthermore, also by the state, mainly metabolic, of the individual itself. The hydra reacts to contacts, to light, heat and electrical stimuli or excitations and to various substances in solution. Such stimuli combine their effects, and through them the creature behaves in accordance with the state in which it finds itself. In green hydra a very marked phototropism is observable, and individuals congregate in the better illuminated spots, wending their way slowly as a rule towards likely foodstuffs and if these latter happen to be microbes or small creatures they first paralyse them by means of nematocysts protruding through the cnidoblasts. Immediately these creatures, like other prey, are carried to the stoma by the tentacles and devoured through movements of the hypostome. In the digestive cavity peristaltic movements take place and the corresponding digestive juices are secreted. Digestion is completed in a way comparable to what takes place in the stomach and gut of a carnivorous animal.

Reactional complexes determined by the behaviour of the coelenterata are better exhibited in more differentiated and larger species, such as certain medusae, the *Gonionemus*. These respond moderately to simple contact and chemical agents such as acids, but the reactions are more intense when such stimuli are added to each other, and particularly when the stimulus approaches the natural excitor, for instance, contact with a pipette containing meat extract, if a small amount of the food be allowed to reach the creature. A flexion is then produced in the tentacles towards the stoma, whilst the hypostome turns towards the contacted tentacle and the nematocysts are touched off. This combination of experimental stimuli resembles those experienced by the medusa under natural conditions in the presence of a small crustacean, for instance, which may serve it for food, and thus the response to stimulus will be identical in each case.

EXTERNAL AND INTERNAL STIMULI

Activity in an animal does not come about solely from some outside stimulus. *Gonionemus* alters its mode of swimming when it gets hungry, as mentioned by Wagner (1905). Loeb (1895) described the inversion of the oesophageal ciliary motion in the actinia *Metridium* (which normally takes place from inside towards the outside) when the creature comes in contact with feeding matter. Such inversion, from outside inwards, to facilitate food ingestion, does not occur in individuals already sated. Allabach (1905) showed, also, that in these creatures hunger increases sensitivity just as among the protozoa. The tenacles of *Stoichactus*, according to Jennings (1905), give no signs of motions anticipatory of food in well-fed specimens. Allabach noted the same in *Metridium*. Gee in 1913 arrested movements of the tentacles in certain *Cribrinos* by squirting food juices into the digestive cavity. Also, Pearl in 1903 observed variations in motion of planarians according to their state of satiety. Similarly, we could quote countless other instances of a like nature.

From the simplest structures upwards, animals react to life in accordance with their environmental conditions and the conditions obtaining between their own internal organs, particularly those handling food. This becomes all the more evident the more advanced the degree of morphological differentiation and functional specialization. Taking worms as our point of departure, the nervous system makes an important leap forward whilst response to outside stimuli, as well as to internal ones, comes about by means of nerve elements answering to changes outside the cells affecting them and likewise to the state of the cells themselves. Thus every action of the nerves depends on excitation and the conditions in which the corresponding structures of neural make-up are placed. These last conditions are in the main of a metabolic and chemical nature.

Thus it is that throughout the whole scale of the animal kingdom, the same thing is repeated and the behaviour of every creature shows itself equally fit as regards responses to external excitants as to stimuli of internal origin; the behaviour of any creature with respect to its environment and under the force of functional or metabolic conditions in the organs of the creature itself is evidence of this.

FITNESS OF ORGANIC FUNCTIONS

For many years now we have laid stress on the adaptability of the organic functions of digestion, circulation, respiration and so on, to the requirements of the individual. "The iron physiological bond linking together the secretory operation of the whole digestive tract, bringing about the secretion of digestive juices," we wrote, as far back as 1908, "becomes the most important factor for stimulating what comes after it, and this together with the natural effect of food on the amount and properties of the various digestive secretions—or to put it in a single word, 'adaptation' of the secretory machinery to the variable digestive circumstances and requirements—are to be seen equally in regard to the dynamics of the various gastro-intestinal departments. The sensitivity of each section of the digestion and the consequent motor responses are adjusted to functional purposes so as to ensure fulfilment of their physiological task in the best possible manner. Motor coordination obeys the laws of functional economy, which likewise govern vegetative activities as well as conscious acts, so that each class of reaction may be exactly superimposed one on the other."

We have repeatedly stated that all functions—especially the vegetative—develop as if governed by some intelligent power. In these involuntary acts the same thing is to be seen as we have just said takes place in animals, in all their forms, right from the very simplest in which there can be not the slightest glimmer of intelligence. Life in the internal organs is, in this respect, superposable on the life of the lower animals. Organic activity evolves quite unconsciously and automatically, but, like other involuntary actions, it corresponds perfectly to some physiological purpose. This explains why one's thoughts always turn towards the idea of the intervention of archei, supposedly projected on to vegetative activity. Everything, in vital manifestations, is accomplished in the best way possible, and if that were not so, life would become extinct, or indeed might never have become established. The living organism—from all departments of life—does its utmost to survive, to be preserved as an individual and to preserve its species.

This is most evidenced in conscious acts. An act is performed for a particular purpose, "with intent." And yet even the higher inner-vations themselves which, in normal circumstances, become manifest in states of consciousness, are founded upon an unconscious pyscho-physiological substructure. It can be understood, therefore, that many of such acts may, in particular cases, be developed in an entirely automatic manner outside the consciousness.

Higher nervous activity is no exception to the general harmony of existence, which comprises the subjection of all vital activities to actual needs. This is the reason why the objective study of psychology is of such tremendous interest.

OBJECTIVE PSYCHOLOGY

Psychology, the science of neuro-psychic existence, is not concerned only with manifestations of the consciousness. Bergson himself admits, for instance, that psychology may never be referred exclusively to acts of consciousness since the domain of the extra-conscious is enormously wider than that of the conscious.

The arrival of a nervous current at the nerve centres or even a change of state in such centres, mainly through metabolic conditions, will promote a reaction. In the first instance it will be a reflex, or act as a spur on the centres. These evolve the more frequently in the outer reaches of the consciousness. The nervous function here again shows itself to be according to plan, directed towards the accomplishment of its particular purpose. In every neural organization, that is to say between a greater or less number of neurones, according to the nature and rank of the acts, relationships are established between the functions from which arises the due fulfilment of the operation.

In the same way that sensations are perceived and integrated by means of images, neural organizations prior to the conscious act, so also are physiological effects of a motor or other type produced along the corresponding nexi, in the form of motor images which may be identified with sensory images, and thus reflexes and automatic motions develop as if they were voluntary acts.

There are various categories of reflex, from the simplest which do not cross the threshold of the physiological, by way of the instinctive reflexes as classified by Bekhterev and the mimic and symbolic reflexes due to nervous concentration, until in the end we reach the most complex types of personal reflex.

Results of the most diverse nature, obtained by physiological investigation of the functions of the nervous system, serve to prove time and again, that in the higher, as well as the lower, nervous activities, as in so many other existential phenomena, the function bows to necessity and obeys its fate, behaving as if for the precise purpose of attaining a particular end, just in the same way as a known and voluntary act may be performed by action of the conscious mind.

This subjection of the whole nervous life to necessity, including therein the self-same voluntary acts of which we have just spoken, arises from the most intimate recesses of existence. Not only, as we have already said, does it come from the excitation of receptors, but is due also to the physiological state of the neurones. Let us not overlook the fact that a neurone is a cell, subject like any other cell to the conditions which make its existence possible. The first condition of its activity is its metabolic state, the way in which nutritional changes from hunger to repletion follow one another, the effects of various substances, metabolites or hormones. An excellent example is to be found in the effect of the respiratory gases according to their respective concentrations on the breathing centres.

Variations in the chemical composition of the blood which flows to irrigate the centres, and the neurones themselves in such centres, will be factors to be reckoned with in the various functional acts. And we may repeat again that conscious acts also are included. To these local influences, derived from the state of the cells, we may add the impulses arriving at such centres by way of the nerves, which routes tally not solely with the somatic nervous system, but also with the vegetative. In this last case also, the currents spring from internal excitations originated in the organism itself. The interposition of internal material or nerve factors is now universally admitted, and the credit for having brought this fact into prominence is due in an extraordinary degree to Turró.

Knowledge and will are not indispensable factors in purpose. Intentionality precedes them by a long stretch. For, everything in life, and for all we know in the physical world as well, is accomplished as if governed by the appearance and course of the phenomena, knowledge, will and intent.

INNERVATIONS AND CONSCIOUSNESS

However, things being thus, it is also certain that the highest inner-vations are accompanied by consciousness. A subjective world evolves side by side with the objective. An afferent current causes a reflex, but at the same time it also arouses a sensation. Such a process is evidenced simultaneously by subjective phenomena and by objective variations. The mystery, scientifically insoluble, consists in the fact that certain physiological mechanisms residing principally in the cerebral cortex are attended by phenomena which are accessible only to what we have called internal observation.

There is a knowledge of the ego—consciousness—which bears with it also knowledge of something which is not the ego, and consequently knowledge of the universe, and of complete reality starting with the existential reality of the subject itself. From the certainty such subject has of the reality which is it itself—"I think, hence I exist" in Descartes' famous phrase—it deduces the certainty of existence of an exterior reality or world of things, which becomes known precisely through mechanisms identical to those which bring us to a knowledge of subjective existence.

As moreover the subject may bring about or not, consciously or at will, certain determinate acts and in this way cause the appearance of certain foreseeable sensations, the subjective criterion becomes naturally converted into an objective existential criterion, as proclaimed by all shades of philosophy throughout the ages.

It is then to be understood how the spectacle of the mind itself, of the consciousness and of the will of the active subject, able to set things in motion and to inhibit, may have constantly permeated its understanding of the universe, and that explanations founded on the subjective stand-ard will have attained so much greater authority the more difficult were the problems and the more complex the questions under consi-deration. In so far as any awkward phenomenon comes into evidence, the anthropomorphic argument comes to the aid of hesitating reason. From the most primitive savage to the most cultured of men, from the basest and most ignorant to the most learned and stately, it is the road of least resistance which leads us to the explanation that everything occurs in our universe through agents referable to the consciousness and the will resulting from human actions of each member of mankind.

Consciousness allows the planning of existence, the projection in series of the acts constituting such existence, movement to a purpose and development of will. This is the foundation of the subjective world, the most profound faculty within the powers of the soul, and the most decisive of rational behaviour! And if this indeed be true, then it is equally probable that the fate of the universe is determined in very fact by an all-pervading Purpose and Will; an Intent and Will beyond mankind, extrahuman and superhuman, supernatural and above— infinitely above—our own individual consciousness and will.

CONSCIOUSNESS AND INTENT

Biology has proven that every vital manifestation is fulfilled with a view towards the future, for the accomplishment of a purpose which makes life possible. If this be not so, there would be no such thing as life. It has always been held that opportunity, adaptation of reactions, might be an exclusive characteristic of conscious acts; and in mechanisms similar to those of the conscious it was naturally claimed that they were of the same order as that observed in the vital phenomena. Facts now are viewed from the opposite end of our perspective. It is not consciousness any more than it is will, or any other powers similar to them, which determine the rules of life; but the knowledge of the ego, the will implied by cognition are merely special manifestations in the individual of universal principles, from which emanates the power of the human mind to turn and twist and regard itself from within, thus acquiring knowledge essential to, for, and of its existence and power!

* * * * *

SANTIAGO RAMON Y CAJAL 1852–1934

Textura del Sistema Nervioso del Hombre y de los Vertebrados (Structure of the Nervous System in Man and Vertebrates), Madrid: Moya, 1899.

As we have already repeatedly explained, the internal prolongations of the spinal ganglia wind their way backwards and emerge from the hind end of the ganglion, forming the posterior roots of the spinal nerves, and, finally, after entering the spinal cord through the postero-lateral groove, fork into two in the substance of the white matter of the posterior cord, to form two branches, one going upwards and the other downwards. From the radicular stem as well as from both the rising and descending branches there sprout out at right angles an infinite number of collateral threads or fibrils which branch out into terminations reaching into various zones of the grey matter.

The nervous impulse arriving from the ganglion through the dorsal root reaches the spinal cord, where it splits into two currents, upward and downward, of equal or maybe unequal intensity. However, as both the stem and the branches of the sensitive root system start off side by side, the nervous wave will be propagated along these latter in order that it may penetrate the grey matter for transmission to the motor and funicular cells.

The terminal stems or branches will constitute the path of conscious impression as well as of widespreading or remote reflexes. This theory, formulated with more or less clarity by ourselves and adopted by various writers, notably Marquez, who used it in order to explain the physiological make-up of the spinal cord, is in agreement with physiological experiments which show us that the excitational energy necessary to arouse a dispersed and distant reflex is much greater than that needed for the production of a unilateral and limited reflex.

There are grounds for supposing that the bonding substance between the terminal branches of the collateral processes and the body or dendrites of the nerve cells is not so perfect a conductor as the nerve protoplasm, but offers a certain resistance evidenced by delay in the passage of the wave. This explains why the delay in travel of the reflex current becomes greater the larger the number of neurones entering into the chain of transmission.

"Reflex actions" is the name given by physiologists to involuntary movements caused by stimulus of the sensitive nerve-endings. In this field of physiology, as in many others, the new theory of structure of the spinal cord is singularly illuminating, allowing an easy understanding of the course of the current in every type of reflex and of the laws governing it, which laws to a certain extent portray mere consequences of the principles of morphology and connections of the medullo-gangliar neurones.

As far back as 1890, we expounded, subject to these new ideas, the first outlines relative to the course of nerve currents from sensory to motor root, without or with the participation of the commissural and funicular cells. Kölliker, van Gehuchten, Waldeyer and Lenhossek followed along these same lines, amplifying our ideas and submitting others in which all the paths which could be taken by a sensory impression in the production of both reflex and voluntary movements were shown in graphic form.

It is moreover an experimental fact that every sensory impulse reaching the outer surface of the brain at the motor-sensory region, through the specific action of the pyramidal cells, produces a sensation.

There are paths leading the impulses up to the cerebral cortex, and of all the paths along the cerebrospinal axis followed by tactile excitations, warmth and pain, from the muscular and tendinous sense, our only exact knowledge at present is of the path pertaining to the sense of touch. There are also paths carrying impulses up to the cerebellum. All these transmissions towards the cortex, in the cerebrum or cerebellum, are made up of various neurones connected in series.

Clearly, if a sensory nerve current travels along collaterals to a higher or lower point of the spinal cord bulb, it will give rise to reflexes. If on the contrary, it reaches the cerebral cortex, the same sensory current will produce a sensation.

Reproduced by permission of Nicolás Moya

IVAN PETROVICH PAVLOV 1849–1936

Лекціи о Работѣ Главныхъ Пищеварительныхъ Железъ (The Work of the Principal Digestive Glands), St. Petersburg: Priroda, 1897.

We are now better able to appreciate the marvellous precision of the working of the digestive glands, which at any given moment supply exactly what is demanded of them—neither more nor less, neither too much nor too little! We can, moreover, rest assured of one fact unique to glandular activity, namely, that these glands are able to prepare secretions of different composition according to requirements, with greater or less amounts of diastases and different proportions of each when—as in the case of the pancreatic juices for instance—several diastases appear in the same secretion. Other properties of the digestive juices also, in addition to their enzyme content, vary according to the requirements of the functions. If note be taken of the quantity secreted in a given period, say an hour, and also of the composition of the corresponding juice, it will be seen that the higher or lower concentration thereof does not depend exclusively upon the rate of secretion. The most diverse proportions are to be observed in ferments between their fluid content and the richness of the juice; a strong digestive capacity is observable whether the secretion be plentiful or sparse. In one and the same secretion, considerable variation can take place independently in the amounts of the various ferments, which shows that glands, like the pancreas, developing complex chemical actions, can supply, according to circumstances and at different times, one product at one instant and another different one later. What is here stated in regard to the diastases is equally true of the amount of fluid, salts and acids present in the secretions.

One important physiological question is to ascertain if these variations in composition of the secretions take place according to digestive requirements for the better development thereof.

A great many facts tell us that the glands are adapted in their working to the various and successive stages of treatment of the food taken in. In the first place, different foodstuffs arouse different intensities of secretion in the organs traversed by them along the digestive tube, for foods are complex and extremely varied in composition. A suitable quantity of each digestive juice will be secreted in succession according to what the meal was made up of.

The working of the glands seems to obey some objective which has to be fulfilled in the best possible way. For instance, vegetable proteins and bread require a great deal of ferment for their digestion. This requirement is met less by an increase in the amount of secreted fluid than by an extraordinary concentration of diastase. It may thus be deduced that what is needed here is an increased amount of ferment, whilst a proportionate concentration of hydrochloric acid might be useless or even harmful. During the digestion of bread, in fact, the presence of an excess hydrochloric concentration in the stomach is usually avoided. The total amount of gastric juice secreted for digesting bread is only slightly greater than that produced for the digestion of milk. The secretion is distributed, nevertheless, over a fairly long period, so that the hourly amount secreted, in the case of bread, comes out at one-and-a-half times less than that observed when dealing with milk or meat. The consequence of this is the low hydrochloric content of the juice for dealing with bread. All of which is in accordance with what biochemistry has to tell us, namely, that the digestion of starch, so plentiful in bread, would be hindered by an excess of acid. A succession of hospital cases shows us quite clearly that with excess gastric acidity, a large portion of the starch fails to undergo digestion and appears therefore in undigested form in the faeces, whilst meat is digested without difficulty.

All these facts and many more show that the working of the digestive glands is extremely complex and flexible, taking place with surprising precision and with the designed purpose of fulfilling a definite physiological need.

SIR CHARLES SCOTT SHERRINGTON 1857–1952

Correlation of Reflexes and the Principle of the Common Path, London: British Association Report LXXIV, 1904.

If we regard the nervous system of any higher organism from the broad point of view, a salient feature in its scheme of construction is the following.

At the commencement of every reflex-arc is a receptive neurone extending from the receptive surface to the central nervous organ. This neurone forms the sole avenue which impulses generated at its receptive point can use whithersoever be their destination. This neurone is therefore a path exclusive to the impulses generated at its own receptive point, and other receptive points than its own cannot employ it. A single receptive point may play reflexly upon quite a number of different effector organs. It may be connected through its reflex path with many muscles and glands in many different regions. Yet all its reflex-arcs spring from the one single shank or stem, i.e. from the one afferent neurone which conducts from the receptive point at the periphery into the central nervous organ.

But at the termination of every reflex-arc we find a final neurone, the ultimate conductive link to an effector organ (muscle or gland). This last link in the chain, e.g. the motor neurone, differs obviously in one important respect from the first link of the chain. It does not subserve exclusively impulses generated at one single receptive source, but receives impulses from many receptive sources situate in many and various regions of the body. It is the *sole* path which all impulses, no matter whence they come, must travel if they are to act on the muscle-fibres to which it leads.

Therefore, while the receptive neurones form a private path exclusively serving impulses of one source only, the final or efferent neurone is, so to say, a public path, *common* to impulses arising at any of many sources of reception. A receptive field, e.g. an area of skin, is analysable into receptive points. One and the same effector organ stands in reflex connection not only with many individual receptive points but even with many various receptive *fields*. Reflexes generated in manifold sense-organs can pour their influence into one and the same muscle. Thus a limb-muscle is the *terminus ad quem* of many reflex-arcs arising in many various parts of the body. Its motor-nerve is a path common to all the reflex-arcs which reach that muscle.

Reflex-arcs show, therefore, the general features that the initial neurone of each is a *private* path exclusively belonging to a single receptive point (or small group of points); and that finally the arcs embouch into a path leading to an effector organ; and that their final path is common to all receptive points wheresoever they may lie in the body, so long as they have connection with the effector organ in question. Before finally converging upon the motor neurone the arcs converge to some degree. Their private paths embouch upon *internuncial* paths common in various degree to groups of private paths. The terminal path may, to distinguish it from internuncial common paths, be called *the final common path*. The motor-nerve to a muscle is a collection of final common paths.

Certain consequences result from this arrangement. One of these seems the preclusion of essential qualitative difference between nerve-impulses arising in different afferent nerves. If two conductors have a tract in common, there can hardly be essential qualitative difference between their modes of conduction; and the final common paths must be capable of responding with different rhythms which different conductors impress upon it. It must be to a certain degree aperiodic. If its discharge be a rhythmic process, as from many considerations it appears to be, the frequency of its own rhythm must be capable of being at least as high as that of the highest frequency of any of the afferent arcs that play upon it; and it must be able also to reproduce the characters of the slowest. Baglioni's results support this inference.

A second consequence is that each receptor being dependent for final communication with its effector organ upon a path not exclusively its own but common to it with certain other receptors, such nexus necessitates successive and not simultaneous use of the common path by various receptors using it to *different or opposed* effect. When two receptors are stimulated simultaneously, each of the receptors tending to evoke reflex action that for its end-effect employs the same final common path but employs it in a different way from the other, one reflex appears without the other. The result is *this* reflex or *that* reflex, but not the two together. Excitation of the central end of the afferent root of the eighth or seventh cervical nerve of the monkey evokes reflexly in the same individual animal sometimes flexion at elbow, sometimes extension. If the excitation be preceded by excitation of the first thoracic root the result is usually extension; if preceded by excitation of the sixth cervical root it is usually flexion. Yet though the same root may thus be made to evoke reflex contraction of the flexors or of the extensors, it does not, in my experience, evoke contraction in both flexors and extensors in the same reflex-response. Of the two reflexes on extensors and flexors respectively, either the one or the other results, but not the two together. Thus, in my experience, excitation of the seventh or eight root never causes simultaneously with reflex contraction of the flexors of elbow a contraction of that part of the triceps which extends the elbow. The flexor-reflex when it occurs seems therefore to exclude the extensor-reflex, and vice versa. If there resulted a compromise between the two reflexes, so that each reflex had a share in the resultant, the compound would be an action which was neither the appropriate flexion nor the appropriate extension. Were there to occur at the final common path algebraical summation of the influence exerted on it by two opposed receptive arcs, there would result in the effector organ an action adapted to neither and useless for the purposes of either.

In the Coelenterate, *Carmarina*, a mechanical stimulus applied to the sub-umbrella causes, as in another Geryonid, *Tiaropsis indicans*, a reflex movement that brings the free end of the manubrium to the spot touched. Bethe reports that if two stimuli are applied simultaneously to opposite points of the discoid sub-umbrella, the points chosen being such that the manubrium is midway between them, the manubrium is moved toward the point at which the stimulus applied was the stronger. He adds that if both stimuli are of exactly equal strength the manubrium remains unmoved and uncontracted. To obtain such a result as this last with antagonistic spinal reflexes in the vertebrate would obviously be more difficult, because the more complex the preparation and the nervous system involved, the more difficult it will be at any moment to balance exactly the two reflexes. But, apart from that, the observation on *Carmarina* is an analogue of that in the monkey's arm.

This dilemma between reflexes would seem to be a problem of frequent recurrence in reflex co-ordination. We note an orderly sequence of actions in the movement of animals, even in cases where every observer admits that the co-ordination is merely reflex. We see one act succeed another without confusion. Yet, tracing this sequence to its external causes, we recognize that the usual thing in nature is not for one exciting stimulus to begin immediately after another ceases, but for an array of environmental agents acting concurrently on the animal at any moment to exhibit correlative change in regard to it, so that one or other group of them becomes—generally by increase in intensity— temporarily prepotent. Thus there dominates now this group, now that group in turn. It may happen that one stimulus ceases coincidently as another begins, but as a rule one stimulus *overlaps* another in regard to time. Thus each reflex breaks in upon a condition of relative equilibrium, which latter is itself reflex. In the simultaneous correlation of reflexes some reflexes combine harmoniously, being reactions that mutually reinforce. These may be termed *allied reflexes*, and the neural arcs which they employ *allied arcs*. On the other hand, some reflexes, as mentioned above, are antagonistic one to another and incompatible. These do not mutually reinforce, but stand to each other in inhibitory relation. One of them inhibits the other, or a whole group of others. These reflexes may in regard to one another be termed *antagonistic;* and the reflex or group of reflexes which succeeds in inhibiting its opponents may be termed "prepotent" for the time being.

The nervous system functions as a whole. Physiological and histo- logical analysis finds it connected throughout its whole extent. Donald- son opens his description of it with the remark: "A group of nerve- cells disconnected from other nerve-tissues of the body, as muscles and glands are disconnected from each other, would be without physio- logical significance." A reflex action, even in a "spinal animal" where the solidarity of the nervous system has been so trenchantly mutilated, is always in fact a reaction conditioned not by one reflex-arc but by many. A reflex detached from the general nervous condition is hardly realizable. The compounding together of reflexes is therefore a main problem in nervous co-ordination. For this problem it is important to recognize a feature in the architecture of the grey-centred (synaptic) nervous system which may be termed *the principle of the common path*.

Reproduced by permission of Lady Sherrington and the British Association for the Advancement of Science

VLADIMIR MIKHAILOVICH BEKHTEREV 1857–1927

Объективная Психология (Objective Psychology), St. Petersburg: Rücher, 1912; this version from the German translation—*Objektive Psychologie*, Leipzig: Teubner, 1913

THE FOUNDATIONS OF PSYCHOREFLEXOLOGY

The aspect of psychology with which we propose to deal in the present book has little in common with what has hitherto invariably been the target of research. In objective psychology or, more accurately, psychoreflexology which is the study of mental reflexes to which we intend to devote the present work, the problems of subjective or conscious processes play no part. As we already know, up to the present, psychological manifestations have been held mainly to embrace conscious phenomena. Professor William James starts his *Textbook of Psychology* with the words, "Psychology may best be defined in the words of Professor G. T. Ladd as the science concerned with the perception and description of states of consciousness," and goes on to say, "among states of consciousness we include phenomena such as feelings, desires, cognitions, reasonings, decisions and the like. The interpretation of such phenomena depends of course on our comprehension of the causes and conditions governing their formation and effects, in so far as these may generally be ascertainable." The material of the study of psychology hitherto, and as it still stands at present, constitutes the inner or subjective world which, being accessible only through individual observation, implies necessarily that the only basic method of procedure open to present-day psychology is that of personal observation.

Some writers do indeed introduce into psychology the notion of unconscious states, but they clothe even these with the properties of conscious conditions and treat them to all intents and purposes as latent conscious phenomena. The whole question of unconscious mental processes in modern psychology is a matter of controversy. A summary of numerous works in this field is available in Dr. G. Cesca's book on the existence of unconscious mental conditions, *Ueber die Existenz von unbewussten psychischen Zuständen*, whilst a thorough treatment of this question is further to be found in Lewis's *Problems of Life and Mind*, J. Mill's *Analysis of the Phenomena of the Human Mind*, Hamilton's *Lectures on Metaphysics and Logic*, and many others; we will therefore at this point dwell no longer on the matter, except merely to point out that over and above those writers who acknowledge the unconscious mental processes, there is ranged a goodly number of psychologists who entirely exclude the unconscious from the mental world. According to Ziehen, for instance, the yardstick of the mental is that which is presented to our consciousness and nothing more than that. The mental and the conscious, as far as we are concerned, according to this writer, run in double harness; we cannot by any manner of means imagine what an unconscious sensation, notion, or the like really is. We are aware of sensations and notions only in so far as we become conscious of them. So-called unconscious activities are nothing more than past and gone material processes which merely excite a mental or conscious process by delayed action. As a concept the unconscious mental process has no meaning.

Other writers besides Ziehen take this attitude. According to them, we cannot postulate an unconscious mental existence in its literal sense. "Though we oft-times speak of unconscious mental existence, this expression is in general either absolutely lacking in meaning or it is not sufficiently precise."

Experience shows, however, that personal observation does not in the least suffice for proper perception of one's own mental life. One opposite example of how misleading it is to allow oneself to be guided by the subjective processes, even in such manifestations as memory and recollection, is given by the memory researches of H. Ebbinghaus who, experimentally, through his own investigation into mechanical learning by rote, became convinced that mental processes which had once taken place, and then apparently vanished from the unconscious, had in fact actually been in existence the whole time.

On the other hand, it is clear that to subjective psychology the whole field of exploration of the conscious processes in individuals other than oneself is barred, since there is absolutely no suitable and accurate means of investigation at its disposal.

In this case the method of scientific investigation clearly constitutes the relation of the individual ego with the extraneous. That this method does not, however, avail is indeed quite plain (as shown in the present writer's paper on "Objective Psychology and its Object"). Similarity or identification in this connection relates, furthermore, to manifestations of two kinds of consciousness of self, which in a different sense are not comparable, and are perceptible only through internal self-observation where no accurate standards are available. Richet is quite right when he says that "inner self-observation, however powerful a method it be, can be applied only to one field—the field of self-knowledge. Outside this it is barren and full of pitfalls." "The *ego* knows itself, perceives itself; it considers and observes itself, and we must not therefore overstep the boundaries of such ego, a domain at once so vast that much still remains to be created from it, and yet at the same time so narrow that our unassuaged thirst for knowledge strives ever to go still further on." Beyond these limits, however, only science with its rigid methods and accurate instruments and measurements, and its slow but sure intellectual development has the power to go. In short, inner observation can rely only on perception of conscious phenomena. The general properties of living matter remain unexplored; their place is in the field of physics, chemistry and physiology.

From our point of view, the usual definition of psychology as a study of the states or manifestations of the consciousness is quite wrong. Psychology must not in reality be restricted to the consideration of conscious phenomena, but must also investigate so-called unconscious mental phenomena and also even the external vital activities of the organism, so far as these are the expression of its mental activities. In brief, the biological basis of mental activity must also be an object of its study.

Does not this imply then, that our movements, whether they be voluntary or involuntary, intelligent or instinctive, must also form an object of investigation from the point of view of subjective psychology.

Do not changes in respiration and heartbeat, evoked by mental processes, also then become an object of psychological study, especially if we reflect that knowledge thereof shows us the correlation between mental phenomena and our bodily functions? A whole series of investigations on the influence of mental processes on the internal organs and bodily workings in general, as well as the effect of intellectual activity on the functioning of the internal organs, must on that account definitely be considered as one of the tasks of psychology, since knowledge of these facts will give us a clue to the explanation of mental phenomena as such, and allow us to understand the essential conditions of the outward manifestations of those activities of the nerve centres which are known under the name of mental processes. Our knowledge of the neuropsychical processes in other creatures similar to ourselves depends also on such conditions.

The intrinsic notion of psychology needs in general to be considerably extended. From our point of view, psychology is the study of mental life in general, and not merely of its conscious manifestations. The business of psychology consists therefore in the diagnosis of all mental processes in the widest sense of the term, that is to say, of conscious as well as unconscious impressions of mental activity, and in appreciating the external expressions of mental activity so far as these serve to determine the properties and characters of such mental activity; furthermore the biological processes, standing in close and direct relation with the mental processes, must also form an object of psychological study. We must not forget that it is a question not only of knowledge of the individual activities of the mind, but also of the mental life of whole groups (such as crowds, communities, nations, and so forth), as well as of the animal kingdom. Hence the division of psychology into individual, social, national, and comparative racial psychology and so-called animal psychology.

Psychoreflexology, which is our subject here, recognizes the material side of the so-called conscious as well as of the unconscious processes, but looks on the mental processes only in their objective manifestations and does not in the least enter into a consideration of the subjective side of the mental world. It has also to investigate and consider the processes in the brain which are accessible to objective research through the use of the finest and most delicate of physical apparatus. Psychoreflexology does not enquire if a process runs its course consciously or unconsciously, leaving that as a matter for subjective psychology on the ground that no objective signs exist entitling us to call certain processes conscious. As we are considering the objective side exclusively, we cannot answer the question whether a process has taken its course within or without the domain of the consciousness. In any event, all researches in this direction lack any scientific significance, since they are little more than slightly confirmed conjectures.

Thus, as is well known, Auerbach discovered that if a frog deprived of its forebrain had its back irritated by acid, it would try to wipe the acid off with the leg corresponding to the irritated side. Is such action conscious or unconscious?

This question has not been answered even yet. Ziehen does indeed say that there is no reason to assume the existence of mental (that is, conscious) analogous processes just because of this performance. On what, however, is such assumption based? We cannot actually penetrate into the mind of the brainless frog, and if we allow ourselves to be guided by personal observation of complicated reflexes within ourselves, it then becomes clear that we cannot deny a certain consciousness in such reflexes—at least, we cannot exclude it. Ziehen denies a parallel course of mental, i.e. conscious processes even in the case of the complex reflexional movements of a frog with its cerebral hemispheres removed as far back as the optic thalami, when, as has been shown, in its hopping it avoids obstacles casting a dark shadow. We have absolutely no need to go into the grounds for this opinion, since there are but few, unless we may accept the writer's adduced analogy of these movements to the automatic motions of a pianist playing his piano or of a man walking mechanically downstairs. Here, however, the fact is overlooked that the pianist and the man going downstairs can carry out these same movements not only automatically but also entirely consciously, whilst we cannot differentiate conscious movements from mechanical or automatic or, in other words, from unconscious movements of this type.

Reproduced by permission of B. G. Teubner, Leipzig

SIGMUND FREUD 1856–1939

Das Unbewusste (The Unconscious), Vienna: *Zeitschrift für Psychoanalyse*, Vol. III, 1915 and *Gesammelte Schriften*, Vol. V, 1924 of the Internationaler Psychoanalytischer Verlag; this version from the authorized English translation—*The Unconscious*, London: Hogarth, 1925.

Psychoanalysis has taught us that the essence of the process of repression lies, not in abrogating or annihilating the ideational presentation of an instinct, but in withholding it from becoming conscious. We then say of the idea that it is in a state of "unconsciousness," of being not apprehended by the conscious mind, and we can produce convincing proofs to show that unconsciously it can also produce effects, even of a kind that finally penetrate to consciousness. Everything that is repressed must remain unconscious but at the very outset let us state that the repressed does not comprise the whole unconscious. The unconscious has the greater compass: the repressed is a part of the unconscious.

How are we to arrive at a knowledge of the unconscious? It is of course only as something conscious that we know anything of it, after it has undergone transformation or translation into something conscious. The possibility of such translation is a matter of everyday experience in psychoanalytic work. In order to achieve this, it is necessary that the person analysed should overcome certain resistances, the very same as those which at some earlier time placed the material in question under repression by rejecting it from consciousness.

In many quarters our justification is disputed for assuming the existence of an unconscious system in the mind and for employing such an assumption for purposes of scientific work. To this we can reply that our assumption of the existence of the unconscious is *necessary* and *legitimate* and that we possess manifold *proofs* of the existence of the unconscious. It is necessary because the data of consciousness are exceedingly defective; both in healthy and in sick persons mental acts are often in process which can be explained only by presupposing other acts, of which consciousness yields no evidence. These include not only the parapraxes and dreams of healthy persons and everything designated a mental symptom or an obsession in the sick; our most intimate daily experience introduces us to sudden ideas of the source of which we are ignorant and the results of mentation arrived at we know not how. All these conscious acts remain disconnected and unintelligible if we are determined to hold fast to the claim that every single mental act performed within us must be consciously experienced; on the other hand, they fall into a demonstrable connection if we interpolate the unconscious acts that we infer. A gain in meaning and connection, however, is a perfectly justifiable motive, one which may well carry us beyond the limitations of direct experience. When, after this, it appears that the assumption of the unconscious helps us to construct a highly successful practical method, by which we are enabled to exert a useful influence upon the course of conscious processes, this success will have won us an incontrovertible proof of the existence of that which we assumed. We become obliged then to take up the position that it is both untenable and presumptuous to claim that whatever goes on in the mind must be known to consciousness.

We can go further and in support of an unconscious mental state allege that only a small content is embraced by consciousness at any given moment, so that the greater part of what we call conscious knowledge must in any case exist for very considerable periods of time in a condition of latency, that is to say, of unconsciousness, of not being apprehended by the mind. When all our latent memories are taken into consideration it becomes totally incomprehensible how the existence of the unconscious can be gainsaid. We then encounter the objection that these latent recollections can no longer be ascribed as mental processes, but that they correspond to residues of somatic processes from which something mental can once more proceed. The obvious answer to this should be that a latent memory is, on the contrary, indubitably a residuum of a mental process. But it is more important to make clear to our own minds that this objection is based on the identification—not, it is true, explicitly stated but regarded as axiomatic—of conscious and mental. This identification is either a *petitio principii* and begs the question whether all that is mental is also necessarily conscious, or else it is a matter of convention of nomenclature. In this latter case it is of course no more open to refutation than any other convention.

The only question that remains is whether it proves so useful that we must needs adopt it. To this we may reply that the conventional identification of the mental with the conscious is thoroughly impractical. It breaks up all mental continuity, plunges us into the insoluble difficulties of psycho-physical parallelism, is open to the reproach that without any manifest grounds it overestimates the part played by consciousness, and finally it forces us prematurely to retire from the territory of psychological research without being able to offer us any compensation elsewhere. At any rate it is clear that the question— whether the latent states of mental life, whose existence is undeniable, are to be conceived of as unconscious mental states or as physical ones— threatens to resolve itself into a war of words. We shall therefore be better advised to give prominence to what we know with certainty of the nature of these debatable states. Now as far as their physical characteristics are concerned they are totally inaccessible to us: no physiological conception nor chemical process can give us any notion of their nature. On the other hand, we know for certain that they have abundant points of contact with conscious mental processes; on being submitted to a certain method of operation they may be transformed into or replaced by conscious processes, and all the categories which we employ to describe conscious mental acts, such as ideas, purposes, resolutions and so forth, can be applied to them. Indeed, of many of these latent states we have to assert that the only point in which they differ from states which are conscious is just in the lack of consciousness of them. So we shall not hesitate to treat them as objects of physiological research, and that in the most intimate connection with conscious mental acts.

The stubborn denial of a mental quality to latent mental processes may be accounted for by the circumstances that most of the phenomena in question have not been objects of study outside psychoanalysis. Anyone who is ignorant of the facts of pathology, who regards the blunders of normal persons as accidental, and who is content with the old saw that dreams are froth, need only ignore a few more problems of the psychology of consciousness in order to dispense with the assumption of an unconscious mental activity. As it happens, hypnotic experiments and especially post-hypnotic suggestion had demonstrated tangibly even before the time of psychoanalysis the existence and mode of operation of the unconscious in the mind.

The assumption of an unconscious is, moreover, in a further respect, a perfectly *legitimate* one, inasmuch as in postulating it we do not depart a single step from our customary and accepted mode of thinking. By the medium of consciousness each one of us becomes aware of his own states of mind; that another man possesses consciousness is a conclusion drawn by analogy from the utterances and actions we perceive him to make, and it is drawn in order that this behaviour of his may become intelligible to us. . . .

We limited the foregoing discussion to ideas and may now raise a new question, the answer to which must contribute to the elucidation of our theoretical position. We said that there were conscious and unconscious ideas; but are there also unconscious instinctual impulses, emotions and feelings, or are such constructions in this instance devoid of any meaning?

I am indeed of opinion that the antithesis of conscious and unconscious does not hold for instincts. An instinct can never be an object of consciousness—only the idea that represents the instinct. Even in the unconscious, moreover, it can only be represented by the idea. If the instinct did not attach itself an idea or manifest itself as an affective state, we could know nothing of it. Though we do speak of an unconscious repressed instinctual impulse, this is a looseness of phraseology which is quite harmless. We can only mean an instinctual impulse the ideational presentation of which is unconscious, for nothing else comes into consideration.

We should expect the answer to the question about unconscious feelings, emotions and affects to be just as easily given. It is surely of the essence of an emotion that we should feel it, i.e. that it should enter consciousness. So for emotions, feelings and affects to be unconscious would be quite out of the question. But in psychoanalytic practice we are accustomed to speak of unconscious love, hate, anger, etc., and find it impossible to avoid even the strange conjunction, "unconscious consciousness of guilt," or a paradoxical "unconscious anxiety." Is there more meaning in the use of these terms than there is in speaking of "unconscious instincts"?

The two cases are not really on all fours. To begin with it may happen that an affect or an emotion is perceived, but misconstrued. By the repression of its proper presentation it is forced to become connected with another idea, and is now interpreted by consciousness as the expression of this other idea.

It is of especial interest to us to have established the fact that repression can succeed in inhibiting the transformation of an instinctual impulse into affective expression. This shows us that the conscious system normally controls affectivity as well as access to motility; and this enhances the importance of repression, since it shows us that the latter is responsible, not merely when something is withheld from consciousness, but also when affective development and the inauguration of muscular activity is prevented. Conversely, too, we may say that as long as the conscious system controls activity and motility the mental condition of the person in question may be called normal. Nevertheless, there is an unmistakable difference in the relation of the controlling system to the two allied processes of discharge. [Note: Affectivity manifests itself essentially in motor (i.e. secretory and circulatory) discharge resulting in an (internal) alteration of the subject's own body without reference to the outer world; motility in actions designed to effect changes in the outer world.]

Whereas the control of the conscious system over voluntary motility is firmly rooted, regularly withstands the onslaught of neurosis and only breaks down in psychosis, the control of the conscious over affective development is less firmly established. Even in normal life we can recognize that a constant struggle for primacy over affectivity goes on between the two systems, consciousness and preconsciousness, that certain spheres of influence are marked off one from another and that the forces at work tend to mingle.

Reproduced from "The Collected Papers of Sigmund Freud," Vol. IV,
by permission of the publishers, The Hogarth Press, London

RENÉ DESCARTES 1596–1650

De Homine (On Man), Leyden: Officinâ Hackiana, 1664.

Animal spirit (nervous force) is like an extremely fine fluid, or we might perhaps better say, an active flame. It is continuously generated by the heart and ascends to the brain, which acts as a sort of reservoir, passing thence to the nerves which distribute it to the muscles where its effect is shown in contraction or relaxation.

On entering the brain cavities, this force permeates the pores of cerebral matter and thus finds its way to the nerves; the amount of energy flowing through each nerve governs the magnitude of contraction in the muscles to which such nerve leads, and gives rise to movement in the corresponding organs.

This arrangement is similar to the underground piping and fountains in our gardens, wherein, according to the head of the water falling from the tanks, different appliances are set more or less forcibly in motion, and various instruments, whistles, and so forth come into play, in keeping with the way the piping, taking the water to any particular apparatus, is fixed up. The nerves, in fact, may be compared to such pipes, whilst the muscles and tendons correspond to the machines and fountains at the end of the line, and the nervous energy itself to the water driving each movement. This nervous energy springs from the heart just like water wells up from the earth, whilst the brain acts as the central reservoir for it in the body.

In addition, breathing and other similar natural and customary actions in the body mechanism, depending on the flow of energy, are like the movements of a clock or a mill, kept in action by the stream of water.

External objects, by their mere presence alone, act upon the sense organs and determine which way the machine is to go. If we take into consideration that the parts of the brain are arranged in a manner similar to the hydraulic plant to which we have referred above, we have an analogy to go upon. Let us assume that a stranger enters the cavern without knowing, for instance, that hidden levers in the floor connected with taps cause the water to run through particular conduits. Our visitor may, perhaps, start making his way towards Diana's Bath, when the soft music of the Pan pipes will strike his ear, and on attempting to proceed towards the spot whence such pleasant sounds seem to come, he approaches the figure of Neptune who starts waving his trident menacingly at him; or, if, on the contrary, he flies in the opposite direction, some sea monster may suddenly start spouting water in his face, and similarly for any sort of arrangement set up according to the fancy of the engineer planning the job.

When a rational mind, say the hydraulic engineer in the illustration given, is in command, it can at will and by reason of its knowledge of the piping with its respective faucets, fittings and accessories bring about different movements by channelling the water gently or vigorously through to the various appliances. In like manner, the rational mind of man—the *power of reason*—resides principally in the brain, where it plays a part similar to that of our waterworks engineer, by sending energy to the nerves and thereby causing the voluntary movements which take place in the muscles.

We now arrive at a final conclusion. All operations of the bodily mechanism, such as digestion of food, beating of the heart, pulsing in the arteries, feeding and growth of the members, breathing, waking and sleeping, vision, hearing, smell, taste, getting warm, and similar qualities revealed by the organs through the external senses, impressions of ideas through common sense or imagination, retention of such ideas in the memory, internal impulses caused by desire and passion, and lastly, the movements of the limbs, which become so accurately adapted to the purposes put to them through the senses, come from the arrangement of the organs and from no other source whatsoever; just as the working of a watch or the action of an automaton depends on the gears and weights which cause it to go. It is not necessary, therefore, to assume the existence of some archeus, whether vegetative or sensory, or any other principle whatsoever, apart from the energy contained in the blood-stream, excited by the heat of that fire which flames constantly within the heart and is identical to that which consumes inanimate and inorganic bodies, to enter into the maintenance of activity in the organs.

Note. From *The Bridge of Life*, by the present author (Macmillan, New York, 1951).

"Descartes' idea of 'animal spirits' is strange to us—an agent simultaneously spirit and substance. Nevertheless, this concept of animal spirits epitomizes the dilemma we are still in when we try to explain excitation. Is it a material, a chemical influence, or an intangible dynamic discharge? To this very day, we cannot say more than that excitation is a manifestation of life, and that it is a distinguishing and specific property of living matter."

GEORGE BERKELEY 1685–1753

A Treatise concerning the Principles of Human Knowledge, Dublin: Jeremy
  Pepyat, 1710.

It is evident to anyone who takes a survey of the objects of human
knowledge, that they are either ideas (1) actually imprinted on the
senses, or else such as are (2) perceived by attending to the passions and
operations of the mind, or lastly (3) ideas formed by help of memory
and imagination, either compounding, dividing, or barely representing
those originally perceived in the aforesaid ways. By sight I have the
ideas of lights and colours, with their several degrees and variations.
By touch I perceive hard and soft, heat and cold, motion and resistance,
and of all these more and less either as to quantity or degree. Smelling
furnishes me with odours, the palate with tastes, and hearing conveys
sounds to the mind in all their variety of tone and composition. And as
several of these are observed to accompany each other, they come to be
marked by one name, and so to be reputed as one thing. Thus, for
example, a certain colour, taste, smell, figure and consistence, having
been observed to go together, are accounted one distinct thing, signified
by the name "apple." Other collections of ideas constitute a stone, a tree,
a book, and the like sensible things; which, as they are pleasing or
disagreeable, excite the passions of love, hatred, joy, grief, and so forth.

But besides all that endless variety of ideas or objects of knowledge,
there is likewise something which knows or perceives them, and
exercises divers operations, as willing, imagining, remembering about
them. This perceiving, active being is what I call *mind, spirit, soul*, or
myself. By which words, I do not denote any one of my ideas, but a
thing entirely distinct from them wherein they exist, or, which is the
same thing, whereby they are perceived; for the existence of an idea
consists in being perceived. . . .

All our ideas, sensations, notions, or the things which we perceive,
by whatsoever names they may be distinguished, are visibly inactive:
there is nothing of power or agency included in them. So that one
idea or object of thought cannot produce or make any alteration in
another. To be satisfied of the truth of this, there is nothing else
requisite but a bare observation of our ideas. For, since they and every
part of them exist only in the mind, it follows that there is nothing in
them but what is perceived: but whoever shall attend to his ideas,

whether of sense or reflection, will not perceive in them any power or activity; there is, therefore, no such thing contained in them. A little attention will discover to us that the very being of an idea implies passiveness and inertness in it, insomuch that it is impossible for an idea to do anything, or strictly speaking, to be the cause of anything: neither can it be the resemblance or pattern of any active being. Whence it plainly follows that extension, figure, and motion cannot be the cause of our sensations. To say, therefore, that these are the effects of powers resulting from the configuration, number, motion, and size of corpuscles, must certainly be false.

We perceive a continual succession of ideas, some are anew excited, others are changed or totally disappear. There is therefore some cause of these ideas, whereon they depend, and which produces and changes them. That this cause cannot be any quality or idea or combination of ideas, is clear from the preceding section. It must therefore be a substance; but it has been shown that there is no corporeal or material substance: it remains therefore that the cause of ideas is an incorporeal active substance or Spirit.

A spirit is one simple, undivided, active being: as it perceives ideas it is called the *understanding*, and as it produces or otherwise operates about them it is called the *will*. Hence there can be no *idea* formed of a soul or spirit; for all ideas whatever, being passive and inert, they cannot represent unto us, by way of image or likeness, that which acts. A little attention will make it plain to anyone, that to have an idea which shall be like that active principle of motion and change of ideas is absolutely impossible. Such is the nature of *spirit*, or that which acts, that it cannot be of itself perceived, but only by the effects which it produceth. If any man shall doubt of the truth of what is here delivered, let him but reflect and try if he can frame the idea of any power or active being, and whether he hath ideas of two principal powers, marked by the names *will* and *understanding*, distinct from each other as well as from a third idea of substance or being in general, with a relative notion of its supporting or being the subject of the aforesaid powers —which is signified by the same *soul* or *spirit*.

The ideas imprinted on the senses by the Author of Nature are called *real things*; and those excited in the imagination, being less regular, vivid, and constant, are more properly termed *ideas*, or *images* of *things*, which they copy and represent. But then our sensations, be they never so vivid and distinct, are nevertheless ideas, that is, they exist in the mind, or are perceived by it, as truly as the ideas of its own framing.

The ideas of sense are allowed to have more reality in them, that is, to be more strong, orderly, and coherent than the creatures of the mind; but this is no argument that they exist without the mind. They are also less dependent on the spirit, or thinking substance which perceives them, in that they are excited by the will of another and more powerful spirit; yet still they are *ideas*, and certainly no idea, whether faint or strong, can exist otherwise than in a mind perceiving it.

Hence, it is evident that those things which, under the notion of a cause co-operating or concurring to the production of effects, are altogether inexplicable, and run us into great absurdities, may be very naturally explained, and have a proper and obvious use assigned to them, when they are considered only as marks or signs for our information. And it is the searching after and endeavouring to understand those signs instituted by the Author of Nature, that ought to be the employment of the natural philosopher; and not the pretending to explain things by corporeal causes, which doctrine seems to have too much estranged the minds of men from that active principle, that supreme and wise Spirit "in whom we live, move, and have our being."

RAMÓN TURRÓ Y DARDER 1854–1926

Ursprünge der Erkenntniss (Sources of Understanding), Leipzig: *Zeitschrift für Sinnespsychologie*, 1911.

Those who fancy that the exteroceptor nerves possess the property of inducing changes of state in the consciousness and, further, in the perception of the reason for such changes—a thing which does not take place in the action of the nerves of general or organic sensibility—are wont to attribute nonsensical physiological properties to such nerves. Voltaire looked on this anomalous, hypothetical property as a miracle distinguishing sensations from without upon their appearance within the consciousness.

And so, indeed, it would be—were it so! Inductive reasoning has, however, come into existence to the undoing of miracles, and teaches us that there is no certainty that impressions received in the centres of exteroceptor perception may be deductively classed as peculiar in cause and effect thanks to some mysterious force created by objective considerations. The driving force comes ever from within, from the trophic feeling which starts off by considering the sensations merely as signs of the source sustaining them. Moreover, since such signs do not make their appearance save in the presence of the sustaining source, they are attributed to the latter as their true causal condition. Between the trophic sensation, the external sensation, and the gastric sensation, a profound connection is established through experience: if the first of these notes the absence of something, the second declares the fact openly by means of signs, whilst the third confirms that the signs are a true correspondence with fact.

This is where consciousness begins. A person eating knows with what he may appease his hunger and what are the things from the outside world which have precisely the capacity of satiating his wants. Such a great discovery would never have been made, had there not existed within his sensory centres prior impressions of whose origin he is unaware. Still, as he observes, through a thousand and one instances, that physiological satisfaction is lacking to the degree that what will allay his hunger does not excite pleasant, tactile or warm sensation in his mouth, or an olfactory impression in his nostrils, an impression of colour in his eyes, and perhaps of sound in his ears, he finishes up by imagining through some inductive process, the terms of which are pre-established in his mind, that these impressions are not born spontaneously in the senses, but that the reappearance thereof is indissolubly linked to the factor advised by the stomach as present when hunger is extinguished.

It is therefore necessary to attribute to the sensory faculties unity, when one observes how the psychotrophic functions are integrated to those of the external sensibility. We are guilty of undue distortion if we assume that the sensorial centres work independently from the start, for we then think that all images and impressions are referable to causes which of themselves have no objective value. If we notice, moreover, that the ingestion of foods is in all cases adapted to the needs of the organism, and thus, if it is surprising and marvellous, for instance, that a hen seeks out the necessary lime-containing substances at the instant it is about to start producing eggs, none the less astonishing is it that all creatures—including man himself—find means of completing their food intake in accordance with nutritional requirements at each particular instant. If what has been previously mentioned is not taken into consideration, still the fact that so much wisdom and reflection is naturally present will cause us to come out of the difficult situation by attributing all these marvels to a combination of miracles which, in order to give it some name, we shall call instinct. Starting, however, with the idea of psycho-physiological unity of the sensory system of internal trophic processes of acceptance with external receptions, through the operation of trophic experience, the initial and most important operation of intellectual existence, we fill in the blanks of our knowledge and the mystery disappears. Nowadays we have at our disposal standards of comparison hitherto never brought into account, which permit us to approach the most obscure questions from an entirely fresh point of view.

Reproduced by permission of J. A. Barth, Leipzig

MIGUEL DE UNAMUNO 1864-1937

Del Sentimiento Trágico de la Vida (On the Fatefulness of Existence), Buenos Aires: *Colección Austral (Espasa-Calpe)*, 1913.

Of all the spokesmen for Objectivism, not a single one pays—or rather had we better say, does not wish to pay—any attention to the fact that when a man asserts his ego, that is to say his personal consciousness, he postulates Man himself the actual and concrete entity in the flesh, and not some mere figment of his own imagination, and in so propounding Man as such he also propounds Consciousness, for the only consciousness of which we have real knowledge is that of Man himself.

The world is designed for consciousness, or rather such design, such notion of purpose or rather than notion let us say perception, such teleological perception does not come into being unless there is a consciousness. Consciousness and purpose are basically the same thing. If the Sun were possessed of consciousness, it would doubtless think that it existed for the purpose of giving light to the universe, but even more than this would it reflect that the universe existed for the purpose of being illuminated by it, so that it should enjoy enlightening the universe and thereby exist. It would be quite right in thinking so!

All this fateful struggle by man to save himself, this eternal craving for redemption, which made Kant give that counterleap from pure to practical reason already mentioned, is nothing but a struggle by the consciousness. If indeed, as some soulless philosopher has asserted, consciousness is no more than a flash in the pan between two eternities wreathed in obscurity, then nothing is more baleful than the fact of being alive.

Some people may see a suspicion of inconsistency in all I am now about to say, sometimes yearning for eternal life and at other times saying that such life possesses nothing of value. Inconsistency? Indeed I think it is! It is the opposition between my heart saying "Yes" and my head saying "No!" Of course there is inconsistency. Who is there that does not remember the Gospel phrase: "Lord, I believe; help thou mine unbelief!" Of course, there is inconsistency, for it is only upon inconsistencies that we live and thrive. Life is doom, and that doom is perpetual struggle with neither victory nor the expectation thereof! All, all is inconsistency!

As you can see, it is all a matter of emotional values, against which reason itself has no chance, because reasons are nothing but reasons and that is the same thing as saying that they are not even facts. There are, indeed, some people who seem to think only with their brain or whatever the organ may be which is specifically detailed for that purpose, whilst on the other hand there do exist some who do their thinking with their whole heart and soul, with their body, their life's blood and the very marrow of their bones, with lungs, belly and everything else about them even to their whole being and life itself. Those who do their thinking merely with their brain turn out to be mere expounders or "professional" thinkers, and we all know what a "professional" is—just a product of the division of labour!

If a philosopher is not a really sincere man, then still less is he a philosopher, for before everything he is then just a pedant, which is to say a mockery of man. The cultivation of any applied science, be it chemistry, physics, mathematics, philology—and even in these cases only very restrictedly and within extremely narrow limits— may perhaps be a matter for detailed specialization; but Philosophy, like Poetry, must be a work of integration and beauteous harmony, otherwise it becomes nothing else but mock erudition, affectation and pretence.

There is one purpose in all knowledge. Gaining knowledge for the mere sake of knowing is not, say what we will, anything but cheerless sophistry. A thing is learned for some immediate practical purpose or in order to complement what we already know. Even a theory which may seem to us most academic, i.e. of the least practical application to our non-intellectual necessities of existence, answers some requirement which, intellectual though it be, is also in very fact a need, whether it is a ground for economy in thought, or a principle of unity and continuity in the consciousness.

But just as scientific knowledge has its aim in further knowledge, so also the philosophy that we may have to embrace has a further extrinsic purpose, all referring to our fate and to our attitude towards life and the universe. To all this, the most fateful problem of philosophy is that of reconciling intellectual requirements with those of the emotions and of the Will. Here lies the breaking point of every philosophy which endeavours to dispense with the eternal and tragic inconsistency that is the basis of our existence. But, do we all face up to this inconsistency?

There is something which, for want of a better name, we shall call the "Fatefulness of Existence," which cloaks a complete conception of life itself and of the universe, an entire philosophy more or less formalized and conscious, that is a perception which, rather than arising from ideas, itself determines them although clearly such ideas may react upon it and corroborate it. This feeling may be possessed and is possessed not only by individuals, but also by entire nations. Nobody has established that man must naturally be cheerful. Consciousness is a heavy burden. . . .

By permission of Dr. Fernando de Unamuno

HENRI BERGSON 1859–1941

Matière et Mémoire (Matter and Memory), Paris: Presses Universitaires de France, 6th Edn., 1933.

This book affirms the reality of mind and of matter, and seeks to determine the relationship between them by studying one definite aspect, that of memory. It is, therefore, frankly dualistic, though on the other hand, it deals with body and mind in such a way as, we hope, to diminish sufficiently, if not entirely, those theoretical difficulties which have always been attendant on dualism, and which cause it to be held in little honour among philosophers despite its being suggested by direct consciousness and accepted by common sense.

These difficulties arise to a large extent out of the idea, sometimes realist sometime idealist, which mankind takes of matter. Our aim is to show that realism and idealism both overstep the mark and that it is a mistake to confine matter to our idea of it, whilst it is equally wrong to make of it something capable of producing perceptions within us though in itself of a different nature entirely. Matter, in our view, is an aggregate of *images* or pictures, by which term we mean a certain existence something more than what the idealist calls a *representation*, mental picture or perception, and something less than what the realist calls a *thing* or object. Our image is found halfway between object and perception, or if you prefer it, between thing and representation. This idea of matter is merely what common sense gives us. A person unaware of philosophic speculations would be greatly astonished if he were told that an object, which he could see and feel in front of him, existed only in his mind and through his mind and was therefore but a figment of his imagination, or even with greater generality, that it existed solely for a mind, as Berkeley maintained. Such a person would always assert that the object did exist independently of the consciousness perceiving it. On the other hand, we should astonish him quite as much if we told him that the object were quite different from what he saw in it, having neither the colour credited to it by his eye nor the solidity apparently felt by his hand. Colour and solidity are for the layman qualities within the object and not states of mind; they are part and parcel of an existence independent of our own. Thus it is that as far as common sense is concerned, the object exists in its very self and has its own particular qualities. In this way a picture is conjured up, existing through its own existence, through the presence of the object in question with all its intrinsic features.

Images inform the whole of our mental life, but there are different shades of such mental life. Our psychological existence can be developed on different planes, more or less close to the action, according to the degree of our faculty of *attending to life*. What is normally held to be a major complication of the psychological condition appears to us as a major extension of our whole personality which, limited generally by action, becomes widened in so far as such limitation is diluted and then, always indivisibly, stretches over a still wider surface. The mental existence is linked inexorably with its concomitant motor in an indestructible unity.

In the majority of cases the psychological state overruns the cerebral to an immense extent. The activity of the brain determines only a very small portion of such psychological condition, being translated into movement or locomotion. Every thought is accompanied by images, at least in its origin, which are not portrayed in the consciousness without outlining in the form of a contour or tendency the movements which would bring them into form in space. Such complex thought depends in its development on the state of the brain at any particular instant, though that is not the end of it.

The relation between the mental and the cerebral is not constant, nor is it simple. The cerebral state contains a greater or less part of the mental according to how we tend to project our psychological life, outward through action, or to draw it inward in the form of pure knowledge. The generality of the idea consists in a certain activity of the mind, in a movement between action and representation, whence it becomes easy for philosophers to locate the general idea in one of the two extremes, boiling it down into words or vaporizing it in recollections; whilst the idea in reality consists in the passage of the mind between one extreme and the other within an indestructible universe.

An English translation of this work is published by Allen & Unwin, London

ERWIN SCHRÖDINGER *b.* 1887

What is Life? Physical Aspects of the Living Cell, Cambridge: Cambridge University Press, 1948.

According to the evidence put forward in the preceding pages the space-time events in the body of a living being which correspond to the activity of its mind, to its self-conscious or any other actions, are (considering also their complex structure and the accepted statistical explanation of physico-chemistry) if not strictly deterministic at any rate statistico-deterministic. To the physicist I wish to emphasize that in my opinion, and contrary to the opinion upheld in some quarters, *quantum indeterminacy* plays no biologically relevant role in them, except perhaps by enhancing their purely accidental character in such events as meiosis, natural and X-ray-induced mutation, and so on—and this is in any case obvious and well recognized.

For the sake of argument, let me regard this as a fact, as I believe every unbiased biologist would, if there were not the well-known, unpleasant feeling about "declaring oneself to be a pure mechanism." For it is deemed to contradict Free Will as warranted by direct introspection.

But immediate experiences in themselves, however various and disparate they be, are logically incapable of contradicting each other.

So let us see whether we cannot draw the correct, non-contradictory conclusion from the following two premises:

(i) My body functions as a pure mechanism according to the Laws of Nature.

(ii) Yet I know, by incontrovertible direct experience, that I am directing its motions, of which I foresee the effects, that may be fateful and all-important, in which case I feel and take full responsibility for them.

The only possible inference from these two facts is, I think, that I—I in the widest meaning of the word, that is to say, every conscious mind that has ever said or felt "I"—am the person, if any, who controls the "motion of the atoms" according to the Laws of Nature.

Within a cultural milieu where certain conceptions (which once had or still have a wider meaning amongst other peoples) have been limited and specialized, it is daring to give to this conclusion the simple wording that it requires. In Christian terminology to say: "Hence I am God Almighty" sounds both blasphemous and lunatic. But please disregard these connotations for the moment and consider whether the above inference is not the closest a biologist can get to proving God and immortality at one stroke.

In itself, the insight is not new. The earliest records to my knowledge date back some 2,500 years or more. From the early great Upanishads the recognition ATHMAN = BRAHMAN (the personal self equals the omnipresent, all-comprehending eternal self) was in Indian thought considered, far from being blasphemous, to represent the quintessence of deepest insight into the happenings of the world. The striving of all the scholars of Vedanta was, after having learnt to pronounce with their lips, really to assimilate in their minds this grandest of all thoughts.

Again, the mystics of many centuries, independently, yet in perfect harmony with each other (somewhat like the particles in an ideal gas) have described, each of them, the unique experience of his or her life in terms that can be condensed in the phrase: DEUS FACTUS SUM (I have become God).

To western ideology the thought has remained a stranger, in spite of Schopenhauer and others who stood for it and in spite of those true lovers who, as they look into each other's eyes, become aware that their thought and their joy are *numerically* one—not merely similar or identical; but they, as a rule, are emotionally too busy to indulge in clear thinking, in which respect they very much resemble the mystic.

Allow me a few further comments. Consciousness is never experienced in the plural, only in the singular. Even in the pathological cases of split consciousness or double personality the two persons alternate, they are never manifest simultaneously. In a dream we do perform several characters at the same time, but not indiscriminately: we *are* one of them; in him we act and speak directly, while we often eagerly await the answer or response of another person, unaware of the fact that it is we who control his movements and his speech just as much as our own.

How does the idea of plurality (so emphatically opposed by the Upanishad writers) arise at all? Consciousness finds itself intimately connected with, and dependent on, the physical state of a limited region of matter, the body. (Consider the changes of mind during the development of the body, as puberty, ageing, dotage, etc., or consider the effects of fever, intoxication, narcosis, lesion of the brain and so on.) Now, there is a great plurality of similar bodies. Hence the pluralization of consciousness or minds seems a very suggestive hypothesis. Probably all simple ingenuous people, as well as the great majority of western philosophers, have accepted it.

It leads almost immediately to the invention of souls, as many as there are bodies, and to the question whether they are mortal as the body is or whether they are immortal and capable of existing by themselves. The former alternative is distasteful, while the latter frankly forgets, ignores or disowns the facts upon which the plurality hypothesis rests. Much sillier questions have been asked: Do animals also have souls? It has even been questioned whether women, or only men, have souls.

Such consequences, even if only tentative, must make us suspicious of the plurality hypothesis, which is common to all official Western creeds. Are we not inclining to much greater nonsense, if in discarding their gross superstitions we retain their naïve idea of plurality of souls, but "remedy" it by declaring the souls to be perishable, to be annihilated with the respective bodies?

The only possible alternative is simply to keep to the immediate experience that consciousness is a singular of which the plural is unknown; that there *is* only one thing and that what seems to be a plurality, is merely a series of different aspects of this one thing, produced by a deception (the Indian MAJA); the same illusion is produced in a gallery of mirrors, and in the same way Gaurisankar and Mount Everest turned out to be the same peak seen from different valleys.

There are, of course, elaborate ghost-stories fixed in our minds to hamper our acceptance of such simple recognition. For instance, it has been said that there is a tree there outside my window, but I do not really see the tree. By some cunning device of which only the initial, relatively simple steps are explored, the real tree throws an image of itself into my consciousness, and that is what I perceive. If you stand by my side and look at the same tree, the latter manages to throw an image into your soul as well. I see my tree and you see yours (remarkably like mine), and what the tree in itself is we do not know. For this extravagance Kant is responsible. In the order of ideas which regards consciousness as a unique principle it is conveniently replaced by the statement that there is obviously only *one* tree and all the image business is a ghost-story.

Yet each of us has the undisputable impression that the sum total of his own experience and memory forms a unit, quite distinct from that of any other person. He refers to it as "I." *What is this "I"?*

If you analyse it closely you will, I think, find that it is just a little bit more than a collection of single data (experiences and memories), namely the canvas *upon which* they are collected. And you will, on close introspection, find that, what you really mean by "I," is that ground-stuff upon which they are collected. You may come to a distant country, lose sight of all your friends, may all but forget them; you acquire new friends, you share life with them as intensely as you ever did with your old ones. Less and less important will become the fact that, while living your new life, you still recollect the old one. "The youth that was I," you may come to speak of him in the third person, indeed the protagonist of the novel you are reading is probably nearer to your heart, certainly more intensely alive and better known to you. Yet there has been no intermediate break, no death. And even if a skilled hypnotist succeeded in blotting out entirely all your earlier reminiscences, you would not find that he had killed *you*. In no case is there a loss of personal existence to deplore.

Nor will there ever be.

CHAPTER XVI

THE WHOLE AND ITS PARTS

CORRELATION AND OBJECT OF MORBOSE REACTIONS

OBSERVATION of the diseased in particular, and of living organisms in general, brings the conviction that there is a correlation of reactions composing life and that such unison is more or less normal or more or less unsettled according to circumstances. This was the view held by the Ancients, notably Hippocrates (467–357 B.C.) and Aristotle (384–322 B.C.). The body is a complete entity responding to the variable conditions of existence in an ordered and harmonious manner as a synergetic whole whose activities are directed to the preservation of life.

To explain processes developing in the best possible way in the living body, throughout the ages, recourse is taken nevertheless to the anthropomorphic theory of directive principles endowed with the human properties appertaining to the subject himself—that is man, the curious and contemplative.

ANALYSIS OF REALITY

What we have to deal with then, is a metaphysical activity which has nothing whatsoever to do with positive science. Still, the whole stage of rational judgment changes when we try to discover what the Universe really is, by considering it as a system based on cause and effect. It then becomes necessary to split things up, to separate out our problems and to study facts by subjecting them to detailed analysis. Descartes (1637) laid down the following principle: "It is wise to split each difficulty under examination into as many separate parts as possible and convenient, in order the better to solve it." Analysis is the great discovery due to Descartes. "The purpose of analysis is to find out by means of one single truth or a particular fact the principles from which it derives." "Analysis is the means of establishing the truth of the first principles of all knowledge." Roger Bacon and Francis Bacon were the famous forerunners of Descartes.

FUNCTIONS OF THE ORGANS

Thus, in order to delve into the mysteries of life, it becomes necessary to study the working of every single one of the various parts of the organism, and of each one of its organs. Observing how processes evolve in the body, and bringing about by experiment the reappearance of such occurrences, we learn to associate certain causes with certain effects, and thus come to establish the experimental method as a systematic and scientific instrument which is the auspicious fruit of positive observation.

Almost contemporary with Descartes were the great anatomists of the Renaissance. Descartes' *De Homine* (*On Man*) came out in 1664, but Vesalius had already published his *De Humani Corporis Fabrica* (*Workshop of the Human Body*) in 1543, not restricting himself merely to morphological descriptions of the organs but investigating also, after the style of Galen, the "uses of the parts." He describes analytically the functions of the muscles, nerves, respiratory system and so on.

Harvey (1628) reduced the problem of the circulation of the blood to a problem in mechanics. And similarly, many others. Everything in life becomes a mechanism. Francis Bacon in 1620 asserts that experiment is of fundamental importance if we want to acquire scientific knowledge, since experiment enables us to establish the causes determining an occurrence, and to bring about such occurrence when we wish. Harvey frequently had recourse to Bacon and applied in practice a rigorous method based on strictly experimental data, allowing him to discover with remarkable accuracy many noteworthy physiological facts which have later been confirmed at the end of subsequent great advances in more recent times.

Experimental physiology and everything of practical value in biology dates from that time. Four centuries of labour, classifying the body into parts, studying the morphological peculiarities of each item, and their separate ways of working, is our present heritage. 'Twas a harsh battle between the experimentalists and the vitalists, a battle which became even more harsh with the rise of positivism in the nineteenth century. Last century, as Haldane pointed out in 1922, young physiologists broke away from vitalist traditions and undertook the investigation of the functions of living bodies, piece by piece, with the hope of arriving thus at a knowledge of the whole working of so complex a mechanism. These mechanicists, however, not only demolished vitalism but at the same time they did away with the idea of life itself.

Progress attained through application of analytical methods to the study of life has been stupendous. Another light on biological processes was lit by this method, as rich and valuable in theoretical learning as in practical application. Modern advances in medicine, physiology and biology are due indeed to the experimental method.

It is worthwhile to observe the functions of the various organs and to cause modifications in their development which may be the consequence of changes imposed in the conditions in which such events take place. The supremacy of causal determinism on vital phenomena becomes evident. Claude Bernard spread this view in the course of many highly important writings but particularly in his *Introduction to the Study of Experimental Medicine* (1865).

Organ by organ, tissue by tissue, cell by cell, the facts of life were explored. The field is bounded by space—in organ, tissue and cell—and by time between the moment a cause commences to act and the instant when its effect is produced, and phenomena are conditioned therefore according to determinism.

Still more is achieved; it becomes possible for cells, and tissues, and eventually the organs also, to continue in existence long after they have been removed from a living creature. From this we come to the conclusion that the parts act and work on their own—like the individual parts of a machine. The whole, whether it be an individual plant or animal, is nothing more in the end than the sum of its parts living in community. The problems of life are solved by analysis, that is, by observation and experiment on the properties of the separate organs.

FUNCTIONAL REGULATION AND CO-ORDINATION

There are of course certain specialized, particular and local functions worthy of individual study. The truth is, nevertheless, that these do not compose the whole of life, which latter is something fuller and more comprehensive than the use of each one by itself and than the summation of individual parts.

In among the common herd of experimenters, who brew up the hotchpot of science, we find a few thinkers of greater vision who warn us, in the face of general opinion, that life also contains other phenomena of wider range than the particularized functions of the organs.

Our attention is first called to the existence of complex mechanisms embracing the functional effect of various organs and then we start the study of inter-organic physiology, the physiology of correlation. Claude Bernard himself, when referring in 1867 to the internal medium in animals, the significance and importance whereof he placed on record, shows that such internal medium must inevitably keep its properties and composition uniform, effecting this by means of complex physiological adjustments which entail many diverse organs; processes of "continuous and delicate balancing like the equilibrium obtaining in the most sensitive of chemical balances" (1878). It was Bernard also who when reporting on the Annual Award of the Paris Academy, claimed by Cyon (1866) on the ground of his discovery of depressor reflexes by cardioaortic receptions, pointed out the importance of the visceral regulative reflexes, by whose intervention functional links became set up among the various evidences of activity in the organs.

Fresh instances keep coming to light, which are characteristic of physiology in the present century, as indicated in our book *Ten Years of Physiology in the Twentieth Century*, published in 1911. The results of countless researches and lengthy thought upon this subject are set forth in two other works, by the present writer, namely *The Unity of the Functions* (1918) and *The Mechanics of Physiological Correlation* (1920).

Of late years there has been an abundant flow of relevant material, and a number of important books have been published. Physiological treatises now devote many a page to discussion of the interfunctional processes, which previously had been dismissed with scarcely a passing comment. Cybernetics, the irrepressible study of control mechanisms, arises from the idea of communication, self-regulation, co-ordination, and physiological adjustment between organs, the collective and the individual psyche. The whole face of physiology has changed in modern times.

PARTIAL AND TOTAL INTEGRATION

Now if this is true in physiology, all the more so do biologists need to make thorough investigations of vital phenomena. The naturalist observes occurrences developing naturally to their ultimate extent; the physiologist carries out experiments, setting down certain conditions and finding the consequences thereof. His judgment has to be determinist, since its source lies in certain causes leading to certain effects. He contemplates fact by sections and brings his ascertained data into their relevant frameworks. Events present themselves mechanistically and therefore the physiologist is usually a mechanicist. In spite of this, during these latter years there has come about, even in the field of physiology, a reversion to unitary ideas of wholeness, totality and fittening of the functions.

The biologist considers extended series of events which develop without his direct participation. The chain of causation does not thrust itself forward, whilst the synthesis, which stands for life and the opportunity for such events to take place, is exhibited without experimental distortion, but in a precisely natural and factual manner. Since the field under survey is so vast, problems of biology seem to be more difficult than those posed by physiology, and it is logical that the naturalist should have a greater leaning towards vitalism. This standpoint and that of the physiologist are two distinct criteriological positions explaining the diametrical opposition of their different opinions.

In all ages, biologists studied the reactions of live bodies as if these were complete, that is to say something more than mere isolated mechanisms within the organs. Roux (1881), Whitman (1888), Driesch (from 1893 onwards), and many others, always supported the unitary principle, and it is safe to say that in every vital manifestation, particularly in ontogenetic growth for instance, as previously mentioned in this book, certain cellular and local properties coincide with certain other general properties belonging to the organism as a whole. During progressive multiplication of the zygote by the increasing number of blastomeres, certain automatic differentiations come into operation due to which each cell element of the embryo evolves on its own according to its nature and intrinsic properties and at the same time dependent differentiations also take place through correlation between parts and regulatory processes, as previously mentioned. Embryogeny continues its development by balancing particular tendencies against general ones. Special inductor, organizatory, evocative, and differentiatory stimulants are necessary to morphogeny, but they must work at any and every instant in agreement with the essential "vital" properties, inherent in each cell and in the whole which is the overall combination of cells.

Form, structure and function make an organization, as Boerhaave wrote, and an organization is something distinct and something more than the mere sum of its separate parts. It possesses properties intrinsic to it as a whole which are not the mere result of a simple addition of the properties of separate elements. The whole exists as a whole and is a real entity quite as evident as any of its factors.

HOLISM

Starting with Smuts (1925), and Mayer (1934) it has become the custom to call general responses *holistic*, in contrast to *meristic* local responses. *Holism*, as it is now called, has recently acquired a certain notoriety, being considered by some as an undefined principle—a brand-new archeus—arising out of the general reactions of organisms, say one of those many theoretical "principles" which time and again have been proposed under the illusion that mere wordy invention of a figment is sufficient to explain a fact. Similar in nature are Monakoff's "hormé," the "libido," and the old-time "neurine," and the rest of the bunch. Unremittingly, time and again whenever events become complex and experimental, determinist analysis becomes difficult there breaks into the field of science some theory of markedly vitalistic colour and distinctive nature.

The "holistic principle" postulates regular working of the organs in the individual and extends even far beyond their direct action. From their presence and activity we should likewise derive the products of human labour in general: civilization, the state, society, culture, arts and sciences. And, according to Mayer, even the very laws of the physical world are subject to the holistic principle, which thus becomes a type of divinity of universal range governing everything and from which nothing can ever escape.

Were this so, then physical laws would have to be deduced from biological concepts and not the reverse as has always been supposed. That would mean coming from complex concepts downwards to the simpler and thence to the simplest, whence the universe would become a holistic aggregate of superimposed totalities each set up upon those below to form a general synthesis—a wholeness of wholenesses. Thence we should get a series of relative entireties ever increasing in their turn, each one complete as far as relates to those of lower rank but being at the same time merely one single part of the next higher. Relations between totalities add up to a combination of all types of force in the universe, which together with the material elements set in motion by such forces make up the substrate of the universe.

Holism, according to Leeman (1935), pertains to a concept of energy whilst existence is an overlaying and synergy of interdependent dynamic fields arranged along the rungs of a ladder of totalities.

The basis of these fanciful speculations is well known; it is co-ordination of the functions, vital unification in the functions and in morphogeny, and fundamentally metabolism. This is the truth as verified by facts, though the conclusions at which we arrive from this starting-point may perhaps be absurd—that depends on any particular writer, and on his followers.

Strict observation of reality teaches us nothing if not that life in the organs develops with its own specific peculiarities, dependent never-theless upon conditions imposed on and by the organism as a whole with its own particular characteristics in its turn. Thus we learn that the functions are related, added together or crossed in their paths, making up something more than a mere simple addition sum of their parts and forming therefore an organization with everything that the word organization connotes and bears in its train; we learn that the functions develop as for the set purpose of a better fulfilment of themselves and the preservation of life by means of ever increasingly complex mechanisms, perfected and efficient in their action. Such observation tells us also that the organism is a unit—the individual—wherein local and general responses take place according to stimuli, with more or less extensive help from different parts, whence, that is from general and local reactions one by one, we get the characteristics of the individuality itself.

ORGANISMAL BIOLOGY

The biology which takes account of general events and their teleo-logical manifestations has come to be called *organismal biology*. An organismal biology must simultaneously cover causal determinant relationships—which admits no discussion—and more general events which seem at times to be sheer probabilities or contingencies likely to elude the actual possibilities of determinist explanation. It is claimed that such a biology might bring mechanicists and vitalists on to common ground, though, up to the present, organismic synthesis has not been able to dispense with hypothetical conjectures of agency of more or less mysterious principles in the course of biological events, leading thus to still another aspect of vitalism seeking now to become re-furbished and brought up to date.

Mechanicism and vitalism are two standpoints of contemplation of the universe, depending on the field of vision, from whence problems are considered, according to the temperament of the student himself so that it becomes difficult to reconcile opinions which are antagonistic from their very roots up.

The man of science does not have to demand from events more than they can offer as means of knowledge. Theories, the tools of the sciences, are needful and aid investigation, but they never attain the validity of factual explanations confirmed by practical experiment and experience. It is absolutely safe to say that physiology took on a new complexion at the start of the present century; that the organism must be looked on as a whole with its parts working together harmoniously as a compact entity and in a fitting manner, this being the distinguishing feature between the events of life and those of the purely physical world. This, however, being shown and granted, it still remains somewhat extravagant to state outright that fresh facts give the lie to the reality and governance of a biological causality determining the operations of the living organism.

BIOLOGY OF THE PARTS AND BIOLOGY OF THE WHOLE

Biology daily discovers fresh prospects and marvellous new domains, wherein a tremendous amount still remains unexplored. Up to the present time, biological investigation has been analytical, studying vital events piecemeal and in isolation. We now know that this dry "meristic" standard means little, and that we must also consider life as a whole, using empirical methods to analyse events as presented naturally by observation, thus bringing reality home to us in its natural state, free from our own interference and a consequent liability to unwitting modification of true conditions. This does not by any means signify that the experimental method is bankrupt, but only that nowadays we have brought before us vaster, more complex and more difficult problems. The extent and deepness of the task need not dismay us, for the fact that these aspects do vary shows us life by one means or another in its manifold and protean forms. In all these cases we shall have to employ the only means of investigation of the universe available to man, our senses and reasoning, our powers of natural or stimulated observation and the logical arrangement of what we have observed.

Whenever a problem has bristled with difficulties, human intelligence has attempted to solve it at the first shot by attributing the observed facts to the agency of some active principle or principles, projected by the spiritual faculties of mankind—knowledge and will —which are thus satisfied with a metaphysical explanation and not a scientific answer. In the second stage man critically observes the facts and discovers a few causal relationships between events; then the mystery is cleared up and the principle is conjured away. Thus step by step, with all the attendant difficulties, we continue to build up our scientific knowledge, though this does not imply that any science can dispense with the metaphysical atmosphere which stands at the peak of all. Hour by hour, man pierces further into the Universe by dint of blood and tears. With every progress, the immensity of the unknown, whose compass we shall never fully know, becomes more and more evident.

At some not too distant date the flamboyant holistic principle will go the same way as its predecessors. This will happen when the events on which the theory is founded can be interpreted scientifically and not metaphysically by establishing what the conditions are which determine such happenings. It is true that these questions are of immense amplitude and their analysis is correspondingly of extreme difficulty, though this will show in the end that as the parts function so does the whole. Life is unity, even though its events are of such widespread variety; nevertheless this life in any of its multifarious appearances is ruled by simple and regular laws. There is nothing extraordinary in the fact that the particular activities of the organs operate one by one, being organized anatomically and physiologically, and that there may be unity in vital responses to external or internal stimulations, and further, that the functions answer a set purpose, being directed to the maintenance of the life of the individual and of the species. To have arrived at these essential and general conclusions, which fill out to such a degree the panorama of events proper to life and which will lighten future tasks, is the masterstroke of contemporary biology!

*   *   *   *   *

JOSÉ DE LETAMENDI DE MANJARÉS 1828–97

La Medicina y el Principio Individualista o Unitario (Medicine and the Unitary or Individualist Principle), Madrid: Inaugural speech at the Medical Reform Club, *Circulo Medico Reformista*, 1882.

First of all you must bear in mind that the times are not favourable for organic, monistic or individualistic theories; that those of you who rally to my banner will have tremendous battles with physiologists, pathologists, therapeutists, *analysts, particularists, anarchists*, and therefore in order that you shall not be caught unawares, I will from the very start do away with the capital objection, or in better words, the sole one which with many variants will be shot at you regarding the unity of man. "Individual unity," you will be told, "does not exist; man is the result of a combination of parts and a knitting together of forces." Let us consider what this objection is really worth. Right away, regarding its form, this is not an objection, it is not an argument, but a mere unproven statement. As to its essential nature, if we cannot talk of *one* man, even less can we speak of *one* lung. Why? Because the relations of man with the universe are those of *contiguity*, whilst those of the lung are of *continuity*. Still more obviously: in order to separate the individual from the environment he lives in, no cutting instrument is required, whilst in order to remove one of his organs from an individual it is essential to cut it out. Well then, where stands unity? Perhaps in the tissue part of the organ? No! Perhaps in the immediate principle? Still no, for the self-same reason of continuity still applies! Is it then resident in the physico-chemical atom? There is some difficulty in asserting this, because the atom, a metaphysical creation of the human mind, is not an object of which we have any positive knowledge.

Thus it is then, that in the analytical or descending scale from the individual downwards, we have no knowledge of any natural unit other than that of the individual.

Whence then does harm spring? From the corrupt direction of reasoning in the application of analysis to the science of the individual, whereby it has been held that analysis is itself constituted a scientific advance, it so being that in anatomy and physiology analysis alone establishes the means of attaining progress, which latter consists in the enlightenment of the synthetic idea of man, in such manner as nature shows him to us.

On this point we are getting worse every day, every day we withdraw more and more from such synthetic idea; we have come to a complete loss of any individual concept. From the death of Hippocrates until the advent of Cartesian philosophy in the seventeenth century, that is to say for more than two thousand years, this concept was sustained, for better or for worse, more or less clearly in its expression; however, since that time, soul and body being divorced from each other by Descartes, the former remaining the plaything of the erudite and psychologists, whilst the latter became the business of doctors and surgeons, the doctors of the Renaissance who fell into the graceless and dangerous extreme of having to base the whole science of anthropology, or shall we say their whole knowledge of man, solely on the investigation of the body, whilst by dint of dissecting and still further dissection, inquiring and still further inquiry, experimenting and still further experimentation, we today have now arrived at so hopeless an elementalism that the most enthusiastic exclusivists, upholders of analysis without end, are already complaining that we are getting lost in a sea of detail, whilst neither in recent physiological treatises nor in those of general pathology, does there remain even a shadow of any of those things, such as character, synergy, temperament, diatheses and many others, which antiquity conceived as a theoretical and practical expression of individual unity and which in the event that antiquity perceived them well and interpreted them badly it was the duty of us moderns to have them put right, and not to have suppressed so much as a single line of fact, whilst they remained throbbing with reality in our hospitals and clinical wards.

Consider that by preserving the individualistic method of Hippocrates in making a diagnosis, and from such prognosis taking the patient as a single whole, whose morbose manifestations, consilient or symptomatic, however incoherent they might appear, form the empirical basis of the qualification of the illness and an indication of its probable course, medicine might throughout its vast sphere have been able to advance for all time, not in spite of physiological and anatomical progress, which has been so disturbing of ideas in other directions, but by virtue of such progress. And if to this we add that advances in the physical sciences, equally upsetting from the clinical viewpoint so far as they claim to be the fountain-head of medical theory, have produced industrial marvels applicable to investigatory and exploratory techniques, we shall get a full idea of the inestimable benefits medicine owes to modern research.

In short, symptoms which in the time of Hippocrates were taken, so to speak, skin-deep from what very naturally was learnt from the patient, are pursued nowadays by the doctor who sounds for them as if they were pearls and coral buried in the deepest depths of the body.

Now you see the conclusions to be drawn from a dispassionate examination:

1. That medicine, in all those artistic processes to which the organs lend themselves, has prospered by what they contain of the *particular;*

2. That medicine has flourished in everything relating to diagnosis, prognosis and treatment founded on observation and experience over the whole individual in his clinical unity;

3. That medicine in everything relating to the building-up of theory, far from thriving is now in a worse state clinically than what it was in the time of Hippocrates, in so far as the ingenuous and prudent ignorance of this great master, protected by an exquisite sense of unity, has been

replaced by a vain and imprudent analytical rashness, not only basing the science of all mankind on the consideration of a single one of its anatomico-physiological elements but also varying even such ground-work in accordance with the deceptive tunes played by every fresh experiment. Ponder deeply on the importance of such aberration; reflect that in medicine, as in any human business, principles are the core of the matter, and there is not a single clinical problem but what, at the least expected moment, a serious question of principle might arise, for though our organs contain a modicum of independent parts, they hold also much of direct dependence on the whole.

JOHN SCOTT HALDANE 1860–1936

Respiration, Oxford: Clarendon Press, 1922.

About the middle of last century the younger physiologists broke away from the vitalistic traditions which had been handed down to them, and set about to investigate living organisms piece by piece, precisely as they would investigate the working of a complex mechanism. This method seemed to them to promise success, and was popularized by such masters of clear and forceful expression as Huxley. It is still the orthodox method of physiology, but the old confidence in it has steadily diminished in proportion as exact experimental investigation has shown that the various activities of a living organism cannot be interpreted in isolation from one another, since organic regulation dominates them. The keynote of this book is the organic regulation of breathing and its associated phenomena.

The mechanistic theory of life is now outworn and must soon take its place in history as a passing phase in the development of biology. But physiology will not go back to the vitalism which was threatening to strangle it, and from which it escaped last century. The real lesson of the movement from that time will never be lost.

This book belongs to a transition period, but the transition is forward and not backward. My treatment of the subject may possibly be looked on askance in some quarters as reactionary: for I have been largely influenced by the ideas and work of older physiologists. If, however, I have gone backward, it is only to pick up clues which had been temporarily lost; and all of these clues lead forward—forward to a new physiology which embodies what was really implicit in the old.

The leaders of the mechanistic movement of last century got rid of vitalism, but in doing so got rid of life itself. I have tried to paint a picture of the body as alive. Though the picture is imperfect, others will soon paint it more completely. The time has come for a far more clear realization of what life implies. The bondage of biology to the physical sciences has lasted more than half a century. It is now time for biology to take her rightful place as an exact independent science: to speak her own language, and not that of other sciences.

The endeavour to represent the facts of physiology as if they would fit into the general scheme of a mechanistic biology has led, it seems to me, to the present estrangement between physiology and medicine. Since the time of Hippocrates the growth of scientific medicine has in reality been based on the study of the manner in which what he called the "nature" (φύσις) of the living body expresses itself in response to changes in environment, and reasserts itself in face of disturbance and injury. The underlying assumption is that organic regulation and maintenance represent something very real, and that only through the study of it can we recognize and interpret disturbance of health, and effectively aid maintenance or restoration of health. I have endeavoured to return to what seems to me the truly scientific Greek tradition, and to give it a form which is not only consistent with modern science and philosophy, but brings physiology and medicine into that close and special relation indicated by the common etymology of the words "physician" and "physiology."

Most of the investigations specially referred to in the book have been carried out on man. It was only by human experiments that the almost incredible delicacy of the regulation of breathing was discovered; and human experiments have revealed to us in other ways how rough many of the experiments on animals, or on "preparations" from the bodies of animals, have been. Organic regulation, with its all-important relations to practical medicine and surgery, was often entirely overlooked. I hope that the book may contribute towards establishing human physiology in its rightful place, which has been usurped too long by experiments on fragments of frogs and other animals, or on the mere superficial physical and chemical aspects of bodily activity.

CLAUDE BERNARD 1813–78

Leçons sur les Phénomènes de la Vie, communs aux Animaux et aux Végétaux (Lessons on Reactions common to Animals and Plants), Paris: Baillière, 1878.

THE INTERNAL MEDIUM

The stability of the *internal medium* is a primary condition for the freedom and independence of certain living bodies in relation to the environment surrounding them. Physiological mechanisms have to function therein assuring the maintenance of conditions necessary for the existence of the cell eleme ts composing them. For we know that there is neither liberty nor independence in the case of the simplest organisms in direct contact with immediate universal circumstances. The possibility of arranging their own internal medium is an exclusive faculty of organisms which have reached a higher stage of complexity and organic differentiation.

Such stability in the internal medium implies an extremely perfect organism, able continuously to balance outside variations. The greater the freedom of the creature with regard to its external environment, the closer will be, on the other hand, the connection of its cells with such internal medium, which will necessarily have to maintain perfect regularity in its qualities, possible only if it has regulatory processes in operation as precise as the most sensitive chemical balances.

The conditions necessary for existence which such internal medium must find will be, first of all, water, oxygen, warmth and food reserves. The same conditions are essential for the continued existence of even the simplest organism. In a polyplastid creature, maintenance of regularity in the qualities of the internal medium are achieved mainly by the intervention of the nervous system, the agency of the functions producing harmony between the activities of the whole of the body. This is attained thanks to properly co-ordinated physiological mechanisms which consolidate the activities of even the most remote organs. To explain these remarkable facts, there is no need to resort to the agency of a vital principle opposed to the influence of physical conditions and possibly peculiar to and characteristic of living organisms. It can be shown that regulatory processes consist exclusively of interfunctional physiological mechanisms.

LUDWIG VON BERTALANFFY *b.* 1901

Theoretische Biologie, Berlin: Bornträger, 1932; this extract from
J. H. Woodger's translation—*Modern Theories of Development: An
Essay on Theoretical Biology*, London: Oxford University Press, 1933

We have made an attempt to think out the implications of the organismic
view in the light of modern physics. It need scarcely be said that, in
view of the extraordinary difficulty and newness of this problem, we
do not wish to defend the foregoing discussion of it in any dogmatic
spirit. Our chief aim has been to draw attention to the problem and to
urge others to investigate it. We would, in any case, point out that as
a *method of investigation* organismic biology is quite independent of
those ultimate decisions which we discussed in the last section. We
cannot indicate this pragmatic attitude better than has been done by
Woodger (1929)—

"If the organism is a hierarchical system with an organization above
the chemical level, then it is clear that it requires investigation at all
levels, and the investigation of one level (e.g. the chemical) cannot
replace that of higher levels. This remains true quite apart from the
remote future possibility of expressing the properties of all higher
levels in terms of the relations between the parts of the lowest level."

Woodger gives an excellent summary of the reasons why an exclusive attention to physico-chemical explanation is not desirable in
biology. To this the reader may be referred for a supplement of what
has been said above. The question whether physical concepts *at present*
suffice for scientific biology must be answered in the negative, because
neurology, experimental embryology, and genetics—to mention only
the more important branches—employ purely biological concepts.
To the question whether these concepts will be replaced by physical
ones in the future, we must answer: wait and see.

It has often been objected against organicism (e.g. Needham, 1929)
that the organismic point of view—although of philosophical value—
is of no importance for the work of natural science. The organism is
something with which the method of natural science cannot work—a
hard, smooth, round nut which experimental analysis can neither crack
nor lever open at any point without it exploding and vanishing like a
Prince Rupert drop. Now, it is quite true that the non-additive
character presents great difficulties to scientific treatment. But

precisely the same difficulties are confronting certain branches of physics —and are here successfully overcome. Köhler (1924), for example, points out the difficulties of a mathematical treatment of the structure of electric charges, since it is impossible to determine the charge first in this place and then in that, because the charge at any given place depends upon that at all the others. Consequently, with the usual additive methods of physics we cannot deal with the problem. The problem must be solved at one stroke as a *whole*, and this physics has done in an admirable way by means of the theory of integral equations. "No one who has closely studied this part of mathematical physics will ever assert that all physical structures have a purely additive character." It cannot therefore be said that the "concept of organism" is opposed to scientific treatment. On the contrary we might say that physics has already been dealing with "organisms," with *Gestalten*, although of a low degree of complication. All that remains of this criticism is the assertion that biology has not yet regarded the organism as a system (as contrasted with an aggregate), and that this is forbidden also for the future. But this is a dogmatism on the side of mechanism which is no better than that of vitalism, when the latter declares that science will "never" be able to explain this or that property of living things.

We can therefore summarize the demands of organismic biology as follows—

Since the fundamental character of the living thing is its organization, the customary investigation of the single parts and processes, even the most thorough physico-chemical analysis, cannot provide a complete explanation of the vital phenomena. This investigation gives us no information about the co-ordination of the parts and processes in the complicated system of the living whole which constitutes the essential "nature" of the organism, and by which the reactions in the organism are distinguished from those in the test-tube. But no reason has been brought forward for supposing that the organization of the parts and the mutual adjustments of the vital processes cannot be treated as scientific problems. Thus, the chief task of biology must be to discover the laws of biological systems to which the ingredient parts and processes are subordinate. *We regard this as the fundamental problem for modern biology.* Since these laws cannot yet be formulated in physical and chemical terms, we are entitled to a biological formulation of them. In our view, the question of a final reducibility of such biological laws is of subordinate importance in view of the foregoing demand. Even without this final decision, the antithesis between mechanism and vitalism ceases to be a troublesome problem. The mechanist who believes in the possibility of such a reduction, and the vitalist who denies it, can join forces in an attempt to solve this great problem: the establishment of the laws of biological systems.

The investigation of these laws must proceed in two directions. On the one hand, the empirical rules of organic systems must be obtained from the concrete, especially experimental, data. And on the other hand, it must be the final aim of biology to derive the laws of organisms deductively from general assumptions.

Reproduced by permission of Prof. Bertalanffy and the Oxford University Press

HANS DRIESCH 1867–1941

Philosophie des Organischen, Leipzig: Engelmann, 1921; this extract based on *The Science and Philosophy of the Organism*, London: Black, 1929.

THE CONCEPT OF ENTIRETY AND INDIVIDUALITY

Whatever may be the theory of *personality*, one thing is certain in practice, namely that there are *distinct* psycho-physical entities. What is the true *essence* of such entities? Or, to put it another way, what is the real living nucleus of a psycho-physical entity? Let us examine this question in greater detail.

A dog "possesses" paws, eyes, coat, hair, and so on; an oak-tree "possesses" leaves. *What* exactly is the "possessor" in these cases? The usual answer to that is of course "the dog," or "the oak-tree" as the case may be. What, however, *is* the *dog* or the *oak-tree*? Could it be the dog's body, or the oak-tree's trunk shorn respectively of paws, or leaves? Is it not rather in each case a matter of entirety which becomes incomplete the moment its paws or leaves are taken away? Yet, what do we mean, in this context, by these words "entirety" and "incomplete"?

Now other complications come into the picture. Suppose instead of a dog, we take a common or garden worm. If we cut this worm's head off, it will grow again. Who "possesses" this power of regeneration, or better still, who "possesses" the power of reproduction of life? Surely not the mere material in the worm's body as such!

We have previously spoken of a deeper *analysis* of the concept of entirety, but must admit right off that such analysis, namely, a precise definition of the concept, is absolutely impossible, since the meaning of "entirety" or of "individuality" is within everybody's reach but only as an outcome of direct intuition.

It is the same thing with many other ideas. If we try to "define" the meaning of "not," "connection," "so many," for instance, we find that definition is possible only if we make use of the meaning of the very thing we are seeking to define; in other words, we can only get a "definition" by going round in circles.

This is what is happening right now with the concept we are in the course of discussing. It occurs time and time again, except in the field of pure logic, where concepts or meanings are dealt with solely in their quality as such in themselves. The reader acquainted with Kant's philosophy will realize that the so-called "deduction" from the category of entirety or individuality can be deduced from the "table of estimates" if the conjunctive estimate $S = P_1 + P_2 + P_3 + \ldots + P_n$ be combined with relative criteria (see my article on this in Vol. XVI of *Kantstudien*). Just the same, I consider Kant's inference to be a type of self-deception, for what is there, pray, in the "table of estimates" which is anything but mere direct intuition?

This digression, however, seems interesting for if, despite everything, we still accept the Kantian thesis, we must reckon also with a possible fresh type of deduction.

We may now say that a concept is "entire" when its essence can be destroyed in proportion to the subtraction of any element from the whole. The concept *A* in such case no longer remains concept *A*, but becomes replaced by a wide variety of *not-A* concepts. Suppose *A* is a "red right-angled triangle"; if we remove the red colour, we are left with only the "right-angled triangle" idea which is no longer the original *A* concept, but a new one, *B*. This holds as far as relates to *logical individuality* or *entirety*.

Let us now look at the same question of individuality or entirety from an objective viewpoint, in the domain of *things* or *events*, which is empirical reality. Things and events are not ideas, but as designated *through* concepts, even when we evade any metaphysical interference.

We know that a table, a steam-engine, a dog, a tree, and the rest form an "entirety." Can this be taken as a definition? Again, we might say, in a purely logical sense, that an "entirety" is something whereof the *essence* distinguishing it would be destroyed were a part of it removed. What portion or how many parts have to be taken away

from a dog for it to stop being a whole? Say a paw! All right, then;
but now consider the dog's coat. If you shave it completely, you can
perhaps say that the doggy essence has been destroyed because its
essential feature is the very fact of having a coat. Now suppose you
cut off three hairs only! Or, if you prefer it, shave off everything *but*
three hairs! How many hairs must you remove before the dog ceases
to be a "whole"?

In the earthworm, it will be impossible for us to say that be-
heading it destroys its wholeness, because the head still exists *latent*
in the de-capitated portion of the body and can be restored through
regeneration.

We must therefore be satisfied with saying that it is through some
kind of instinct that we know what objective entirety may be. This is,
however, not enough because some objective standard of entirety still
seems to be needed.

We shall get closer to the solution if we say a thing or an event is
complete when countless cases or copies of it exist, each one with the
same typical shape, which can be used as a standard of comparison
against any particular example we possess. Such an answer is still not
perfect, since works of art, for instance, may be complete though there
is usually only one specimen of each.

We might also say: "an entirety" or "complete whole" is what gets
restored when its material structure has been disturbed in some way or
other. There are indeed many organisms able to regenerate, as well as
artificial products of human activity, industry and art, which can be put
together again. Nevertheless cases frequently occur also of living
creatures, mammals and others, which though obviously forming an
entirety are incapable of repairing any damage they have suffered.

In my opinion, the only valid standard of entirety or individuality
would be if we said that "entirety" is something owing its origin to
non-mechanical natural agents and is at the same time something
rather more than a mere "sum" of its parts. This standard implies that
non-mechanical natural agents tend to entirety through their own
conditions.

"Entirety" however is not the only idea we need for logical understanding of organism. We cannot do without *combination* or *union* of
the ideas of *entirety* and *causality*, the discussion whereof will lead us to
examine the important question which may be considered as the
keystone for a logical justification of our vitalism, namely *whether different
elemental forms of cause and effect are logically possible*, or whether causality

must always be mechanical, even though we use the word "mechanical" in its widest sense.

For, if causality were wider than summative or mechanist causality, our theory of free will in life would have received *some definite logical justification.*

This is what we are after. Biological concepts have to be subjected to *logic*, taking this term in its most general sense of a "theory of order," allowing us to conceive empirical reality as "one great ordered whole," wherein every singularity of being or occurrence occupies its own precisely determined right place. In this way, we should realize what I have termed a "monism of arrangement," which will nevertheless continue to be but an ideal, since the structure of reality has two aspects, a blend of causal *entirety* and fortuitous *non-wholeness.*

In any event, it is reasonable to ask oneself whether inanimate, non-living nature could also partake of entirety. Henderson, in his famous book *The Fitness of Environment*, revealed clear pointers on this when he observed the "*fitness* of the inorganic world for the development and existence of life"; in other words, the pre-established harmony between existence and environment.

Let us say, in conclusion, that our present knowledge of entirety and individuality is still a bit shaky, since we lack a stable and suitable standard of judgment. Only personal organism, fortunately, forms an exception. *The biology of the person is the prototype of the doctrine of objective entirety and individuality, accurately individualizing causality and evolution.*

Reproduced by permission of A. and C. Black, Ltd., London

JOSÉ ORTEGA Y GASSET b. 1883

Towards a Philosophy of History, New York: Norton, 1941.

Human life is a strange reality concerning which the first thing to be said is that it is the basic reality, in the sense that to it we must refer all others, since all others, effective or presumptive, must in one way or another appear within it.

The most trivial and at the same time the most important note in human life is that man has no choice but has to be always doing something to keep himself in existence. Life is given to us; we do not give it to ourselves, rather we find ourselves in it, suddenly and without knowing how. But the life which is given us is not given us ready-made; we must make it for ourselves, each one his own. Life is a task. And the weightiest aspect of these tasks in which life consists is not the necessity of performing them but, in a sense, the opposite: I mean that we find ourselves always under compulsion to do something but never, strictly speaking, under compulsion to do something in particular, that there is not imposed on us this or that task as there is imposed on the star its course or on the stone its gravitation. Each individual before doing anything must decide for himself and at his own risk what he is going to do. But this decision is impossible unless one possesses certain convictions concerning the nature of things around one, the nature of other men, of oneself. Only in the light of such convictions can one prefer one act to another, can one, in short, live.

It follows that man must ever be grounded on some belief, and that the structure of his life will depend primordially on the beliefs on which he is grounded; and further that the most decisive changes in humanity are changes of belief, the intensifying or weakening of beliefs. The diagnosis of any human existence, whether of an individual, a people, or an age, must begin by establishing the repertory of its convictions. For always in living one sets out from certain convictions. They are the ground beneath our feet, and it is for this reason we say that man is grounded on them. It is man's beliefs that truly constitute his state. I have spoken of them as a repertory to indicate that the plurality of beliefs on which an individual, a people, or an age is grounded never possesses a completely logical articulation, that is to say, does not form a system of ideas such as, for example, a philosophy constitutes or aims at constituting. The beliefs that coexist in any human life, sustaining, impelling, and directing it, are on occasion incongruous, contradictory, at the least confused. Be it noted that all these qualifications attach to beliefs in so far as they partake of ideas. But it is erroneous to define belief as an idea. Once an idea has been thought it has exhausted its role and its consistency. The individual, moreover, may think whatever the whim suggests to him, and even many things against his whim. Thoughts arise in the mind spontaneously, without will or deliberation on our part and without producing any effect whatever on our behaviour. A belief is not merely an idea that is thought, it is an idea in which one also believes. And believing is not an operation of the intellectual mechanism, but a function of the living being as such, the function of guiding his conduct, his performance of his task.

This observation once made, I can now withdraw my previous expression and say that beliefs, a mere incoherent repertory in so far as they are merely ideas, always constitute a system in so far as they are effective beliefs; in other words, that while lacking articulation from the logical or strictly intellectual point of view, they do nonetheless possess a vital articulation, they *function* as beliefs resting one on another, combining with one another to form a whole: in short, that they always present themselves as members of an organism, of a structure. This causes them among other things always to possess their own architecture and to function as a hierarchy. In every human life there are beliefs that are basic, fundamental, radical, and there are others derived from these, upheld by them, and secondary to them. If this observation is supremely trivial, the fault is not mine that with all its triviality it remains of the greatest importance. For should the beliefs by which one lives lack structure, since their number in each individual life is legion there must result a mere pullulation hostile to all idea of order and incomprehensible in consequence.

The fact that we should see them, on the contrary, as endowed with a structure and a hierarchy allows us to penetrate their hidden order and consequently to understand our own life and the life of others, that of today and that of other days.

Thus we may now say that the diagnosing of any human existence, whether of an individual, a people, or an age, must begin by an ordered inventory of its system of convictions, and to this end it must establish before all else which belief is fundamental, decisive, sustaining and breathing life into all the others.

Reproduced by permission of the publisher

MANUEL GARCÍA MORENTE 1888–1944

La Filosofía de Kant (Kant's Philosophy) Madrid: Victoriano Suárez, 1917.

INTERNAL FINALITY

Nevertheless, in the group of things we call organisms, such mechanical explanation cannot be claimed until a most precise and detailed knowledge has previously been obtained first of the form and next of the functions of such organisms. Nor can such knowledge be built up without abundant use of the principle of finality. Indeed, organisms

are chunks of reality formed in such a manner that there is to be discovered within them what Kant calls an internal finality or purpose.
A thing possesses internal finality when it is itself both the cause and the
effect of itself. A tree produces another tree like itself, which is to say,
from the viewpoint of species, a holm-oak, for instance, begets itself,
is its own origin and cause and its own effect or result. However so it
is also from the individual viewpoint, for the oak evolves and develops
its whole concept by internal power; it does not harbour, as a mechanical addition, its material elements but is also nourished on them,
converting them totally and embodying them into its own form. In
the living organism, the preservation of the whole depends on the
preservation and working of its parts, and these in their turn depend on
the survival of the whole. If the addition of fresh elements in live
organisms is not something extra but an assimilation, similarly a taking
of them away is equivalent to death or at least demands that a neighbouring organ shall fulfil by proxy the functions of the organ which
has been done away with.

The parts of a live organism are, therefore, organs, that is to say, that
far from having no concern with the whole, their form and working
must be determined according to the idea of the whole body. The
totality of the organic being is an end, whose maintenance is pur-
posed by the parts or organs. An end, comprised in a concept or idea,
must decide *a priori* everything which has to be contained in it.

That, however, is not enough. Any object of human ingenuity,
a watch for instance, is at the same time a combination where the idea
of the whole determines from beforehand its composition and the
shape of its parts. We do not, because of that, say that the watch
possesses internal finality or purpose. In order that a thing may possess
internal purpose, that is, for it to be an organism, it is necessary in
addition that the complete whole arises in its turn from the shape and
working of its parts. These two requirements appear to contradict one
another. On one hand, the organism in its whole conception determines the special organs and their working, whilst on the other it is the
organs and their operation which beget and preserve the organism.
This contradiction is exactly the principle problem in biology as a
science of existence. Existence or continued life is precisely an internal
end, a system of forms wherein each part is determinate and at the
same time determinant, wherein each part begets the whole and is
in turn begotten in accordance with the design of the whole. An
organism is an organized being which organizes itself. The watch of

our illustration is on the other hand a thing organized by something outside itself.

FINALISM AND MECHANICALISM

The foregoing statement expounds adequately the part played by the principle of finality in the science of live creatures. Internal finality is not an explanation of life, but a specific character of the latter, a character which has to be explained mechanically.

Biology's problem is to give a mechanical explanation of life. Life, however, shows itself to our observing faculties with an internal purpose. Thus, the knowledge we need to acquire from forms (morphology) and from functions (physiology) will be necessarily founded upon the principle of finality; we shall always have to investigate the why and wherefore of such-and-such form and of such-and-such function. Live beings, then, appear before our observation as internal systems of ends and means. If, however, biology be limited to a study of forms and functions—morphology and physiology—it would be nothing more than a mere description of what is and what happens, interpreted in a fashion convenient for mankind, but it would not be scientific knowledge nor would it be an explanation of existence. Biology purposes to find such explanation; a host of modern theories seek to formulate it with more or less success. We do not here have to go into its study. It has been sufficient for us to find the indicative, regulative, auxiliary function which the principle of finality possesses in the study of organisms.

INDEX OF AUTHORS QUOTED

INDEX

(Page numbers printed in italic type refer to quoted extracts.)

327